D1538773

SOLAR SCIENCE

EXPLORING SUNSPOTS, SEASONS, ECLIPSES, AND MORE

Dennis Schatz

Andrew Fraknoi

National Science Teachers Association

Arlington, Virginia

National Science Teachers Association

Claire Reinburg, Director
Wendy Rubin, Managing Editor
Amanda O'Brien, Associate Editor
Rachel Ledbetter, Associate Editor
Donna Yudkin, Book Acquisitions Coordinator

ART AND DESIGN
Will Thomas Jr., Director
Joe Butera, Senior Graphic Designer, cover and
    interior design

PRINTING AND PRODUCTION
Catherine Lorrain, Director

NATIONAL SCIENCE TEACHERS ASSOCIATION
David L. Evans, Executive Director
David Beacom, Publisher

1840 Wilson Blvd., Arlington, VA 22201
www.nsta.org/store
For customer service inquiries, please call 800-277-5300.

www.fsc.org

MIX

Paper from
responsible sources

FSC® C005010

Copyright © 2016 by the National Science Teachers Association.
All rights reserved. Printed in the United States of America.
19 18 17 16   4 3 2 1

*NSTA is committed to publishing material that promotes the best in inquiry-based science education.
However, conditions of actual use may vary, and the safety procedures and practices described in this book
are intended to serve only as a guide. Additional precautionary measures may be required. NSTA and the
authors do not warrant or represent that the procedures and practices in this book meet any safety code
or standard of federal, state, or local regulations. NSTA and the authors disclaim any liability for personal
injury or damage to property arising out of or relating to the use of this book, including any of the recommendations, instructions, or materials contained therein.*

PERMISSIONS
Book purchasers may photocopy, print, or e-mail up to five copies of an NSTA book chapter for personal
use only; this does not include display or promotional use. Elementary, middle, and high school teachers
may reproduce forms, sample documents, and single NSTA book chapters needed for classroom or non-
commercial, professional-development use only. E-book buyers may download files to multiple personal
devices but are prohibited from posting the files to third-party servers or websites, or from passing files to
non-buyers. For additional permission to photocopy or use material electronically from this NSTA Press
book, please contact the Copyright Clearance Center (CCC) (*www.copyright.com*; 978-750-8400). Please
access *www.nsta.org/permissions* for further information about NSTA's rights and permissions policies.

Library of Congress Cataloging-in-Publication Data
Schatz, Dennis, author.
  Solar science : exploring sunspots, seasons, eclipses, and more / Dennis Schatz, Andrew Fraknoi.
    pages cm
  Summary: "Solar Science offers more than three dozen hands-on, inquiry-based activities on many
fascinating aspects of solar astronomy. The activities cover the Sun's motions, the space weather it causes,
the measures of time and seasons in our daily lives, and much more."-- Provided by publisher.
  Includes index.
  ISBN 978-1-941316-07-8 (print) -- ISBN 978-1-941316-47-4 (e-book) 1. Astronomy--Experiments--Juvenile
literature. 2. Sun--Observations--Juvenile literature. 3. Sun--Experiments--Juvenile literature. 4. Sun--
Juvenile literature. I. Fraknoi, Andrew, author. II. Title.
  QB46.S2592 2015
  523.7078--dc23
                                2015031629

e-LCCN: 2015035458

Dedication

To Alan J. Friedman,
good friend, colleague,
and mentor, who inspired
everyone he met to remember
that science is a way of
thinking, not a list of facts.

CONTENTS

Understanding and Tracking the Annual Motion of the Sun and the Seasons

Solar Activity and Space Weather

The Sun, the Moon, and the Earth Together: Phases, Eclipses, and More

About the Authors

Dennis Schatz is the author of numerous resources for educators and museum professionals, including *Astro Adventures: An Upper Elementary Curriculum* (Pacific Science Center 2002) and *Astro Adventures II* (Pacific Science Center 2003). He is also the author of 23 science books for children that have all together sold almost 2 million copies worldwide and have been translated into 23 languages. These include *Astronomy Activity Book* (Simon and Schuster 1991) and *Stars and Planets* (SmartLab Toys 2004).

Dennis was a member of the five-person design team that developed the Earth and space sciences disciplinary core ideas for the National Research Council that are found in *A Framework for K–12 Science Education*, which was used to develop the *Next Generation Science Standards*.

For many years, Dennis was the senior vice president for strategic programs at the Pacific Science Center in Seattle, Washington. For four years he served as a program director for science education at the National Science Foundation. At the Pacific Science Center, he codirected Washington State LASER (Leadership and Assistance for Science Education Reform), a program to implement a quality K–12 science program in all 295 school districts in Washington State. He was also principal investigator for Portal to the Public, an initiative to develop programs—both on-site and off—that engage scientists in working with diverse audiences to enhance the public's understanding of current science research.

He has received numerous honors, including several from the National Science Teachers Association (NSTA): the 2009 Faraday Science Communicator Award, the 2005 Distinguished Service to Science Education Award, the 1996 Distinguished Informal Science Education Award, and the 1980 Ohaus Honorary Award for Innovations in Science Teaching.

More information about Dennis Schatz is available at
www.dennisschatz.org.

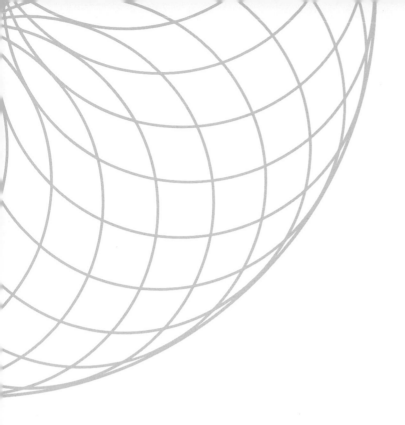

Andrew Fraknoi is the author of *Disney's Wonderful World of Space* (an astronomy book for grades 5–7) and is the lead author of several successful introductory astronomy textbooks for nonscience majors (such as *Voyages Through the Universe*, 3rd ed., published in 2004 by Brooks-Cole/Cengage). In the 1980s, he also edited two books of science

and science fiction for Bantam. He is editor and coauthor of *The Universe at Your Fingertips 2.0*, a collection of astronomy activities and teaching resources published by the Astronomical Society of the Pacific that is in use in formal and informal educational institutions around the world.

He is the chair of the astronomy department at Foothill College, Los Altos Hills, California, and appears regularly on local and national radio explaining astronomical developments in everyday language. Fraknoi was the cofounder and coeditor of *Astronomy Education Review*,

the online journal and magazine published by the American Astronomical Society. The International Astronomical Union has named Asteroid 4859 Asteroid Fraknoi to recognize his contributions to astronomy education and outreach (but he wants us to mention that it's a very boring asteroid and no threat to the Earth!).

Andrew is the winner of the 2012 Faraday Science Communicator Award from NSTA, as well as the 2007 Andrew Gemant Award from the American Institute of Physics. Also in 2007, he was selected as the California Professor of the Year by the Carnegie Foundation for the Advancement of Teaching. His other awards include the Annenberg Foundation Award for astronomy education from the American Astronomical Society and the Klumpke-Roberts Award for public outreach in astronomy from the Astronomical Society of the Pacific.

For more about Andrew Fraknoi, see
www.foothill.edu/ast/fraknoi.php.

Introduction

The Sun is not only the easiest astronomical object in the sky to observe, but it has a greater influence on our lives than any other cosmic object. The fundamental elements of how we mark time—the day and year—are based on the Earth's relationship with the Sun. The cycle of our seasons (from winter to summer and back to winter again) has to do with the tilt of the Earth toward or away from the Sun as a year passes.

The charged particles streaming from the Sun's surface (the solar wind) cause the spectacular auroras that people travel thousands of miles and brave the cold of northern nights to see (Figure I.1). When these particles overload our planet's storage capacity, they can disrupt our radio transmissions and interrupt electrical power distribution. As our civilization gets more and more interconnected on Earth and in space, the chances of serious problems from a storm on the Sun increase.

The unique relationship between sizes and distances in the Earth–Moon–Sun system also cause spectacular eclipses of the Sun and the Moon. Total solar eclipses that reveal the eerie glow of the Sun's outer atmosphere (the corona) are so beautiful and rare that people travel from all over the world to witness them (Figure I.2).

Both the new and the older science standards suggest that students need to have a good fundamental understanding of the Sun's effect on their daily lives. There's no better way of providing

FIGURE I.2

Total solar eclipse showing the Sun's corona, which becomes visible during totality

that understanding than through the kinds of eye-opening (indeed, mind-opening) experiences in this volume.

This book is specifically designed for instructors of grades 5–8 who teach about the Sun and Moon and their cycles as well as eclipses, but some of the materials could easily be adapted for higher grades or (informal) settings outside of school. Throughout, we provide classroom-based activities, background information, and

FIGURE I.1

(*Left*) Aurora (northern lights) above Bear Lake in Alaska

experience with the science practices and crosscutting concepts identified in the *Next Generation Science Standards* (*NGSS*). The core of the book is a series of student-centered learning experiences that put the students in the position of being scientists: asking questions; exploring phenomena; and drawing, discussing, and refining conclusions.

Educators who use the experiences and suggested teaching strategies in this book will involve students in the three-dimensional learning process recommended by the *NGSS*, effectively integrating the teaching of the disciplinary core ideas, science practices, and crosscutting concepts related to the Sun and the Moon and their motion in the sky. Sections strategically located throughout the chapters identify especially good places to emphasize the three-dimensional learning that students are experiencing.

In addition, we provide a variety of resources that connect the content to best practices in mathematics and literacy, including the use of age-appropriate web pages and real-time data from observatories and satellites, such as the Solar Dynamics Observatory (Figure I.3).

Although the first edition of this book is timed to allow teachers to prepare for the Great American Eclipse of 2017, which will be visible throughout the United States, the book is not tied to that event and will be useful for teaching about the Sun and its effects on our culture and our understanding of nature in any year.

FIGURE I.3

A digital rendering of the Solar Dynamics Observatory satellite

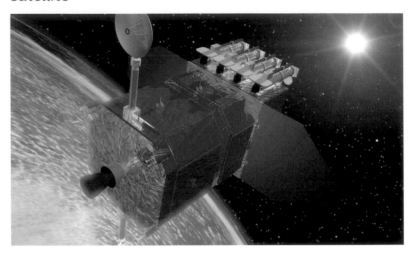

How This Book Is Organized

The book contains four chapters, each of which deals with disciplinary core ideas in the *NGSS*:

1. **Understanding and Tracking the Daily Motion of the Sun:** What does the Sun do in the sky each day, and how does that relate to our notions of time and direction?

2. **Understanding and Tracking the Annual Motion of the Sun and the Seasons:** How does the Sun's motion and position in the sky vary throughout the year, and how does that relate to our ideas of a calendar and the seasons?

3. **Solar Activity and Space Weather:** What phenomena do we observe on the surface and in the atmosphere of the Sun, and how do these influence what we observe and how we live our lives on Earth?

4. **The Sun, the Moon, and the Earth Together: Phases, Eclipses, and More:** How do the relationships among the Earth, the Moon, and the Sun produce solar and lunar eclipses?

Each chapter identifies the specific performance expectations, disciplinary core ideas, science practices, and crosscutting concepts in the *NGSS* that are addressed in the activities. Also listed are connections that the experiences make with the math and literacy standards in the *Common Core State Standards* (*CCSS*).

Learning experiences in each chapter move students from initially engaging with the disciplinary core ideas, science practices, and crosscutting concepts to having a deeper understanding of each of them. These experiences follow the successful 5E Instructional Model developed by Biological Sciences Curriculum Study (BSCS; see the more detailed explanation of the 5E Model, p. xviii), dividing the experiences into the following five categories:

 Engage experiences hook the students into wanting to learn more about the topic and reveal their preconceptions about the subject.

 Explore experiences allow students to build from their preconceptions by making observations (e.g., by viewing the Sun through special glasses [Figure I.4, p. xvi]), and using the scientific practices to generate questions and consider new ideas based on their observations.

 Explain experiences allow the teacher, via continued student discussion and activities, to help students develop a deeper and improved understanding of the core disciplinary ideas, scientific practices, and crosscutting concepts.

 Elaborate experiences provide opportunities for students to apply their new level of understanding to related questions or topics.

 Evaluate experiences allow students (and teachers) to gauge how well they understand the concepts covered in the chapter.

Each learning experience in the book provides all you need to organize, prepare, and implement it, offering the following information:

1. Overall concept: A general description of the experience.

2. Objectives: What students will learn or produce by completing the experience.

3. Materials: Everything you need to have before you begin.

4. Advance preparation: Important steps that need to be completed before you are ready to have the students do the experience.

5. Procedure: Step-by-step directions for your students and you, plus answers to the questions we suggest asking. Also included are alternative approaches to deal with different classroom structures (e.g., a self-contained classroom vs. multiple sections throughout the day).

Finally, each chapter ends with suggestions for activities to connect to math and literacy concepts and other resources that allow for further exploration of the concepts by students and teachers.

We do not expect that every teacher will use every experience in each chapter. The goal is to provide you with a wealth of activities so that you can choose the ones that best fit your students' developmental level, your class structure, and your time limitations. At a minimum, we think it is important you include at least one each of the *engage*, *explore*, and *explain* experiences.

As students will learn using these experiences, what they discover about the Sun's motion across the sky, the phases of the Moon, and other astronomical phenomena will be different for people at different locations on the Earth. Given that this is a publication primarily for use in the Northern Hemisphere, some of the experiences will need modification to be effective at latitudes outside mid-Northern Hemisphere locations (i.e., outside the continental United States and southern Canada).

Use of an Astronomy Lab Notebook

If you already use notebooks or journals in your class, you can skip this explanation of their value, although it is important to understand the difference between notebooks and journals in general and why we refer specifically to using astronomy lab notebooks.

Journals are typically dominated by personal reflections on what the student is currently thinking and how the student is reacting to what is happening in his or her daily life—somewhat akin to a memoir. Science notebooks hold more structured, objective descriptions of science experiments and observations being made by students, plus conclusions reached based on the data collected. While a science notebook also allows for

FIGURE I.4

Students learn about safe viewing of the Sun.

occasional reflection regarding students' initial understanding of a concept and how their thinking changes after analyzing the collected data and discussion with other students and the teacher, it always comes back to making sense of observations and data (see Figure I.5 for an example of Galileo Galilei's use of a notebook).

There are numerous reasons why having students keep astronomy lab notebooks is valuable:

- Students have a single place to document what they have done over an extended period of time and across a number of science experiences. This makes it easy to recover details that students may have forgotten.

- Students can reflect on how their understanding of a concept has changed from the beginning of a unit of study to the end. This type of reflection—especially written reflection—is key to deeper learning (an idea well confirmed by research on how people learn).

FIGURE I.5

Drawings of the Moon from Galileo's astronomy lab notebook

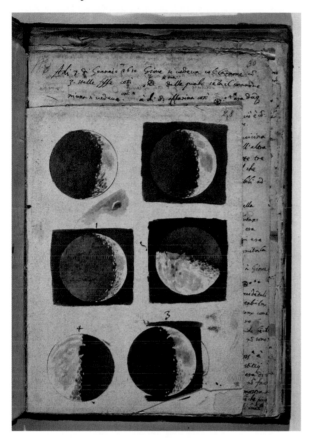

- Teachers can see—and assess, if necessary—what students have done step by step and how well the students understand the concepts being studied.

- It creates an entry point for discussion between student and teacher or among students, which can lead to deeper understanding of the concept being studied.

- Keeping the journal enhances students' writing skills, which support all areas of study by them.

- The process of keeping a contemporaneous lab notebook mirrors what scientists do in their daily lives to document and reflect on their area of research.

Although there is no one way to use notebooks, here are a few criteria that we think are important for creating an astronomy lab notebook:

1. Whatever you use as your notebook (e.g., commercial spiral notebook or composition book; student-created notebooks from 8 ½ × 11 paper), it is important to let the student personalize the notebook including his or her name on the cover along with artwork of an astronomical nature. This can consist of drawings made by the student or astronomical images printed from the internet or cut out of magazines.

2. Several pages at the beginning should be reserved for a table of contents, which is completed as students add material to the notebook.

3. The pages should be numbered so material can be easily referenced in conversations with the teacher and other students.

4. Each experience in our book suggests the minimum material that should be included in the notebooks to document what students are doing. Sometimes this calls for writing directly in the notebook, and other times it involves attaching observations or a worksheet into the notebook. You may wish to have students add more detail than we suggest regarding the procedures they follow or to reflect more on predictions and conclusions.

If you are new to the use of science notebooks, here are some of our favorite resources related to their use:

- *Science Notebooks in Middle School* (from the Full Option Science System program at the Lawrence Hall of Science): *http://goo.gl/glnddU*

- Science Notebooks (Washington State LASER [the website was formerly kept by the North Cascades and Olympic Science Partnership]): *www.wastatelaser.org/Science-Notebooks/home*

- Fulwiler, B. R. 2007. *Writing in science: How to scaffold instruction to support learning.* Portsmouth, NH: Heinemann. *www.heinemann.com/products/E01070.aspx.* (Read a free sample chapter at *www.heinemann.com/shared/onlineresources/e01070/chapter2.pdf.*)

Use of the 5E Instructional Model

The use of a learning cycle method to teaching science goes back to at least the 1960s and is based on a constructivist learning approach that emerged from what research tells us about how people learn (e.g., Bransford, Brown, and Cocking 1999). There are a number of learning cycles used by different curricula, but we have settled on the 5E Instructional Model developed by the BSCS because it is one of the most studied and widely used approaches (see *http://bscs.org/bscs-5e-instructional-model* for more background).

If you already use a learning cycle in your classroom, you can skip the rest of this section.

If not, here are the key attributes to keep in mind during each step of the 5E Model:

ENGAGE

- **What it is time to do:** Hook the students' interest and curiosity to learn more, determine the students' current understanding and preconceptions, and raise questions the students want answered.

- **What it is not time to do:** Provide definitions, conclusions, or the right answer; lecture to students; or explain concepts.

EXPLORE

- **What it is time to do:** Provide opportunities to conduct and record observations, have students work together in research teams to compare results and discuss their data, have students suggest conclusions based on the data, ask probing questions of students that encourage them to reflect on their thinking and redirect their investigations if needed, make sure that everyone (including the teacher) is a good listener, and help all students be actively engaged in collecting and discussing data.

- **What it is not time to do:** Allow students to stop considering other conclusions once a single solution is offered, give students answers or provide detailed explanations of how to work through a problem, tell students they are wrong, or give information or facts that solve the problem immediately.

EXPLAIN

- **What it is time to do:** Encourage students to explain concepts and definitions in their own words, based on justification using data; explain concepts relying on experiments or observations; and build from students' previous experiences and preconceptions.

- **What it is not time to do:** Accept or propose explanations without justification or discourage or stifle students' explanations.

ELABORATE

- **What it is time to do:** Expect students to apply the concepts, skills, and vocabulary they have learned to new situations; make additional observations and collect data; and use previous learning to ask questions, design new experiments or observations, and propose appropriate solutions.

- **What it is not time to do:** Provide students with detailed solutions to new problems, provide answers before thorough discussion, or ignore previous data and evidence related to the concept.

EVALUATE

- **What it is time to do:** Have students demonstrate and assess their own learning regarding the concepts while the teacher assesses student progress.

- **What it is not time to do:** Introduce new concepts or subjects or test for isolated vocabulary and facts.

Teachers across the country have found the 5Es a useful and effective way to proceed through their science lessons. At first, individuals not used to the rhythm or pace of the 5Es might get impatient with them, but research shows that allowing students to learn through their own investigations makes for far deeper and longer-lasting learning than just lecturing to them or showing a video.

Use of Think-Pair-Share

The think-pair-share learning strategy is also encouraged throughout the book and is closely aligned with the 5Es. Like the 5Es, the think-pair-share sequence is based on many years of research regarding how people learn. It provides a mechanism for individual students to personally reflect (think) on a question or topic, typically in writing, before having a discussion with a partner or small group of students (pair). The sharing allows students to further reflect on their own and others' thinking about a subject, often leading to refined and improved understanding. Different groups may initially come up with different ideas or solutions about the problem at hand. When all the groups have a chance to discuss their thinking in front of the entire class (share), a wider range of ideas come to the fore. Guided by a skillful (and not too intrusive) teacher, the class can list, discuss, and evaluate the various ideas the groups have come up with and deepen everyone's understanding.

Personalizing the Use of This Book for Your Class

Although this book and its student handouts can be used effectively on its own, it may also

be supplemented by a textbook, by student research in a library or on the internet, and by audiovisual materials that help reinforce or develop concepts that are hard to visualize. We have listed some of our favorite outside resources throughout the book.

We understand that different classes and groups have different schedules. Some teachers reading this book will have a class for most of each day and can apportion their time in such a way that experiments can be carried out throughout the day. Other teachers will have each class for only one period and—if experiments need to be continued at another time on the same day—may need to pool the work of several classes to obtain the data needed to continue. We have tried to take into account both kinds of classes in the instructions, but we also appreciate that no one knows a class better than its teacher, and thus you may want to modify our suggestions to fit your particular circumstances.

We also know (from our past experience leading professional development workshops for teachers) that the readers of this book will likely find ways to improve, expand, and personalize the experiences we suggest for their students. If you come up with great new ways of treating the subjects we cover or simply find a clever modification of our suggestions, we would love to hear from you. You can reach us by e-mail at *dschatz@pacsci.org* (Dennis) and *fraknoi@fhda.edu* (Andrew).

Both of us were trained as astronomers and have spent our professional careers explaining the wonders of the sky and the greater universe to students, teachers, museum educators, and museum visitors. Nothing gives us greater pleasure than when someone looks up and says, "Oh, I get it! For the first time, I get how that works."

May you have many such experiences with your students as you use this book.

—Dennis Schatz and Andrew Fraknoi

Special Note of Acknowledgment

Student activities, like the experiences in this book, are often put forward by a number of authors independently. One version is often influenced by other adaptations that educators try and then report on in write-ups, conference sessions, and web pages.

We gratefully acknowledge inspiration for sections of a number of these experiences from the talented staff of the following institutions: the Stanford Solar Center; the NASA Science Missions Directorate Heliophysics Forum; the Exploratorium science museum in San Francisco; the University of California, Berkeley, Space Sciences Lab Center for Science Education; Hands On Optics; the Astronomical Society of the Pacific; the Pacific Science Center; David Huestis at Skyscrapers, Inc.; NASA's Goddard Space Flight Center; the Chabot Space and Science Center; the National Oceanic and Atmospheric Administration, NASA's Stratospheric Observatory for Infrared Astronomy (SOFIA), and others. We are particularly grateful to Deborah Scherrer of the Stanford Solar Center for many useful suggestions.

Finally, many thanks go to Paul Allan, coauthor of *Astro-Adventures II* (Pacific Science Center 2003), who developed and refined many of the activities that are the basis of a number of experiences in this book.

A Website for This Book

A special set of web pages to go with this book has been established at the National Science Teachers Association (NSTA) website. You can find it at *www.nsta.org/solarscience*.

The page includes the following:

- PDF versions of student forms, material templates, and handouts for the book

- An updated list of links from the book (so that you don't have to retype URLs)

- News and resources about upcoming eclipses and other topics related to the book

- New ideas and reader suggestions for the experiences in the book

REFERENCE

Bransford, J. D., A. L. Brown, and R. R. Cocking, eds. 1999. *How people learn: Brain, mind, experience, and school*. Washington, DC: National Academies Press.

Flares on the Sun

1

Understanding and Tracking the Daily Motion of the Sun

The apparent movement of the Sun across the sky is one of the fundamental facts of life on our planet. The daily motion of the Sun is not only a key concept identified in most state standards, but it also provides the perfect opportunity for students to collect their own scientific data for analysis. This leads them to develop their science practice skills by analyzing and interpreting their data. They then make meaning of their observations to explain how the Sun moves and how we measure time. They can further develop their skills by building and using physical and mental models to describe the observed apparent motion of the Sun across the sky each day (Figure 1.1).

FIGURE 1.1

An illustration of the Sun's path across the sky in different seasons

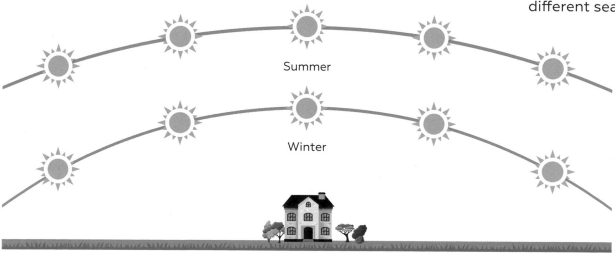

Summer

Winter

Learning Goals of the Chapter

After doing these activities, students will understand the following:

1. The Sun moves from lower in the eastern part of the sky in the morning to higher overhead during the middle of the day to lower in the western sky in the afternoon.

2. This motion of the Sun is due to the Earth rotating on its axis and not to the Sun actually moving across the sky.

3. The Sun is highest in the sky at local noon but is not generally directly overhead in the United States.

4. The height of the Sun above the horizon at any time of day depends on your location on the Earth.

5. The position of the Sun in the sky, and the shadow this produces of objects, can be used to tell time or to identify the cardinal directions (north, south, east, and west).

6. The time of day at a given moment depends on your location on the Earth. We established time zones around the Earth to manage measuring time.

Overview of Student Experiences

Teacher note: Not all the listed experiences need to be done to have a successful learning sequence. We provide a wide set of experiences so that you can tailor the unit to your students' needs.

ENGAGE EXPERIENCES

These are designed to hook the students into wanting to learn more about the changing position of the Sun across the sky during the day and to explore how this can be monitored by the shadows produced by the Sun's light.

- **1.1. Cast Away: What Do We Think We Know?** Students first consider being stranded on a tropical island with Chuck (Tom Hanks) in the movie *Cast Away*. They brainstorm all the ways they could measure time, from short periods, like seconds, to longer periods, like years. This discussion leads students to consider and record their thoughts regarding the following questions: (1) How does the Sun's position in the sky change throughout the day, and what causes that change in position? (2) How different is the motion of the Sun in the sky for people at other locations, both elsewhere in the United States and at other locations outside North America? (3) If you have a cousin who lives in England, what systems have people created so that you can arrange a specific time to talk on the phone?

- **1.2. Your Personal Pocket Sun Clock:** Students construct a portable Sun clock to use at school, at home, or anywhere in between, and then explore how the clock needs to be oriented in the sunlight to give the correct time. They begin developing the concept that time is related to the changing position of the Sun in the sky throughout the day.

EXPLORE EXPERIENCE

This experience has students collect and analyze data about the Sun's position and the shadows it produces to build the background needed to understand what emerged from the *engage* experiences and to prepare them to develop and explore the physical and mental models in the Explain section of the chapter.

- **1.3. Shadow and Sun Tracking:** Students make simple sundials with poster paper and toilet plungers as gnomons (pronounced *NO-mans*) and record the direction and length of the plunger's shadow at various times during a single school day. Students also observe the location of the Sun relative to local features (e.g., buildings, flagpoles, trees) at least three different times throughout the day. The data are added to their astronomy lab notebooks, and writing prompts ask them to discuss the relationship between the direction and height of the Sun above the horizon and the direction and height of the gnomon's shadow on their sundials.

EXPLAIN EXPERIENCES

Students use the data collected in the prior experiences to develop and use physical and mental models that help them understand what they have observed about the Sun's daily motion across the sky.

- **1.4. Modeling the Sun–Earth Relationship:** Students model the Sun–Earth relationship using their heads as the Earth and a bare lightbulb at the front of the room as the Sun to understand what causes the Sun to appear to move across the sky. They use this model to explain some of the basic concepts in the chapter, such as the meanings of noon, midnight, sunrise, and sunset.

- **1.5. Noontime Around the World:** Students use toothpicks and balls of clay to make gnomons for their globes to mark their location on an Earth globe. A bare lightbulb a few feet away provides a Sun for the Sun–Earth model. By rotating the globe, students see the same change in shadow length and direction that they saw in their own observations outside. Other gnomons can be added to the globe to see how the shadows vary for different locations and allows the students to explore the difference in time for other locations on Earth.

ELABORATE EXPERIENCES

Students engage in additional activities, as time allows, to reinforce and practice using the physical and mental models developed in previous experiences.

- **1.6. Pocket Sun Compass:** Students experiment with their pocket Sun clock to determine how it can be used as a Sun compass and to reinforce how the Sun clock works, based on the mental and physical model developed through the *explain* experiences.

- **1.7. High Noon:** Students re-examine the gnomon shadow data they collected earlier to identify when the Sun's shadow was the shortest. They then use simple geometry and the law of similar triangles to determine the height (in degrees) of the Sun above the horizon at noon, thus emphasizing that the Sun is not always directly overhead at noon—a preconception that many people have.

EVALUATE EXPERIENCES

These experiences allow students to demonstrate their mastery of the science practices developed in the chapter. We would not expect teachers to do all of the *evaluate* experiences but to choose the ones that best fit the learning goals for their students and the integration strategies they themselves use. (For example, the first experience connects well with language arts skills desired in students.)

- **1.8. Write a Picture Book for Kids:** Each student writes and illustrates a children's picture book for a younger student to explain how the Sun moves across the sky during the day. Writing prompts are provided to encourage the inclusion of key ideas learned in the chapter.

- **1.9. Where Is It Night When We Have Noon?** Students use the equipment from "Noontime Around the World" to complete a diagram about where it is nighttime and midnight when it is noon at different places on the Earth, including their home location.

- **1.10. What Do We Think We Know? Revisited:** Students revisit their responses to the questions from the *engage* experience to discuss how their thinking has changed after completing the activities in the chapter.

Recommended Teaching Time for Each Experience

Table 1.1 provides information about the teaching time needed for each experience in the chapter. Please note that many of the experiences require sunny days. These are indicated in the table.

Connecting With Standards

Table 1.2 (pp. 6–7) shows the standards covered by the experiences in this chapter. This book does not deal with aspects of the *Next Generation Science Standards* performance expectations that involve the stars at night. Chapter 2 deals with any phenomena associated with seasons and the seasonal changes related to the apparent motion of the Sun in the sky. Chapter 4 addresses lunar phases and eclipses. Additionally, this table does not give specific *Common Core State Standards* but gives general connections to the language arts and mathematics standards.

TABLE 1.1

Recommended teaching time for each experience

Experience	Time
Engage experiences	
1.1. Cast Away: What Do We Think We Know?	**60 minutes**
1.2. Your Personal Pocket Sun Clock ●	**30 minutes** to assemble; **45 minutes** for first observation; **10 to 15 minutes each for two** follow-up observations; **20 minutes** for follow-up discussion
Explore experiences	
1.3. Shadow and Sun Tracking ●	**60 minutes** for first assembly and first observation; **10 to 15 minutes each for five or six** follow-up observations; **30 minutes** for follow-up discussion
Explain experiences	
1.4. Modeling the Sun–Earth Relationship	**45 minutes**
1.5. Noontime Around the World	**60 minutes**, which it is good to split into two sessions over two days
Elaborate experiences	
1.6. Pocket Sun Compass ●	**40 minutes**
1.7. High Noon ○	**45 to 90 minutes**, depending on the need to get more observations
Evaluate experiences	
1.8. Write a Picture Book for Kids	**120 minutes** over several days, or do a portion as homework
1.9. Where Is It Night When We Have Noon?	**45 minutes**
1.10. What Do We Think We Know? Revisited	**45 minutes**

● = Requires a sunny day   ○ = May require a sunny day

TABLE 1.2

Chapter 1 *Next Generation Science Standards* and *Common Core State Standards* connections

Performance expectations	• 5-ESS1-2: Represent data in graphical displays to reveal patterns of daily changes in the length and direction of shadows, day and night, and the seasonal appearance of some stars in the night sky.
	• MS–ESS1-1: Develop and use a model of the Earth-Sun-Moon system to describe the cyclic patterns of lunar phases, eclipses of the Sun and Moon, and seasons.
Disciplinary core ideas	• 5-ESS1.B: (Earth and the Solar System): The orbits of Earth around the Sun and of the Moon around Earth, together with the rotation of Earth about an axis between its North and South Poles, cause observable patterns. These include day and night; daily changes in the length and direction of shadows; and different positions of the sun, moon, and stars at different times of the day, month, and year.
	• M-ESS1.A: (The Universe and Its Stars): Patterns of the apparent motion of the Sun, the Moon and stars in the sky can be observed, described, predicted and explained with models.
Science and engineering practices	• Develop and use a model to describe phenomena (e.g., a model of the Sun–Earth system).
	• Analyze and interpret data to determine similarities and differences in findings (e.g., analysis and interpretation of shadows produced by the Sun).
	• Engage in argument from evidence (e.g., discussion of how the Sun's shadow varies for different places on Earth).
Crosscutting concepts	• Patterns can be used to identify cause-and-effect relationships (e.g., analysis of the Sun's path across the sky and the resulting shadow of the gnomon on the Sun clock).
	• Science assumes that objects and events in natural systems occur in consistent patterns that are understandable through measurement and observation (e.g., analysis of the Sun's path across the sky and the resulting shadow of the gnomon on the Sun clock).
	• Models can be used to represent systems and their interactions (e.g., use of an Sun–Earth model to understand the Sun–Earth system).

Table 1.2 *(continued)*

Connections to the *Common Core State Standards*	• *Writing:* Students write arguments that support claims with logical reasoning and relevant evidence using accurate, credible sources and demonstrating an understanding of the topic or text. The reasons and evidence are logically organized, including the use of visual displays as appropriate.
	• *Speaking and listening:* Students engage effectively in a range of collaborative discussions (one-on-one, in groups, and teacher led) with diverse partners, building on others' ideas and expressing their own clearly. Report on a topic or text or present an opinion, sequencing ideas logically and using appropriate facts and relevant, descriptive details (including visual displays as appropriate) to support main ideas or themes.
	• *Reading:* Students quote accurately from a text when explaining what the text says and when drawing inferences from the text. Students determine the meaning of general academic and domain-specific words and phrases in a text relevant to the student's grade level.
	• *Mathematics:* Students recognize and use proportional reasoning to solve real-world and mathematical problems. Students summarize numerical data sets in relation to their context, including reporting the number of observations and describing the nature of the attribute under investigation, including how it was measured and its units of measurement

Content Background

EARTH'S ROTATION AND DAY AND NIGHT

The key idea for this chapter is that the Earth spins on its axis once every 24 hours (once each day) from west to east. As a result of this rotation, the Sun, the Moon, the planets, and the stars all appear to move in the sky in the direction opposite to our rotation—from east to west. This is why each day the Sun appears to rise roughly in the east, move across the sky as the day goes on, and set roughly in the west (Figure 1.2, p. 8).

It is the Earth's rotation that defines the length of our days and nights and gives us the most basic rhythm of time on Earth. The day and night cycle is how we organize our lives, and it is the fundamental unit that goes into the making of our calendars, although how we define other time units has always been up to humans to decide. Today we use seconds, minutes, and hours to mark smaller units of time, but this was not always the case. Various ancient cultures subdivided the day in different ways.

FIGURE 1.2

In the course of a day, the Sun rises roughly in the east, moves across the sky, and sets roughly in the west.

FIGURE 1.3

An illustration of the duodecimal counting system

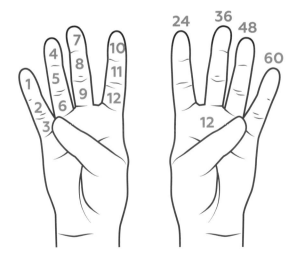

MEASURING TIME

We owe our systems of 12 and 24 hours, 60 minutes, and 60 seconds to the *duodecimal counting system* that was common in Asia and the Middle East thousands of years before the Common Era. Instead of just counting to 10 using all our fingers and then starting again, this system had a more complex counting method that used the three segments into which each of our fingers, except the thumb, is divided. The four fingers with three segments on one hand let you count to 12, using your thumb as a physical device to touch to each segment as you count. The other hand allows you to keep track of the number of 12s you have counted, and its five fingers let you count 5 × 12 or 60 (Figure 1.3).

Before humans had mechanical clocks, we were still able to get a good measure of time over the course of a day using sundials, which have existed for at least 5,000 years. Although

FIGURE 1.4

Three different sundials: (a) a smiley face sundial on the side of a Kaunas University of Technology building, Kaunas, Lithuania; (b) a sundial in Forbidden City, Beijing, China; and (c) a typical style of sundial in a city park

(a)

(b)

(c)

sundials come in many different forms, they all share the common trait of having a gnomon (a protruding stick or narrow edge) that produces a shadow (caused by the Sun's light) on a surface marked by hours of the day (Figure 1.4). Human convention is that we define noon as the time the Sun crosses the line in the sky going from north to south. Thus, the gnomon on a sundial is always aligned in the north–south direction.

TWO KINDS OF NORTH

It is useful to understand the two different definitions of north on our planet. There is celestial (or geographic or rotational) north and south, defined by where the Earth's axis (the line around which our planet rotates) "sticks out" from our planet. There is also magnetic north, defined by where the ends of the Earth's magnetic field come out of the Earth. These two north poles are not the same. In fact, the north magnetic pole tends to move around because

of the complexity of our planet's magnetic field (Figure 1.5, p. 10). The difference in direction of celestial north and magnetic north (as measured as an angle in degrees) is called *magnetic declination*. It varies at different locations (and changes a bit with time as the magnetic north pole moves around).

Compass needles point to magnetic north, whereas the determination of north from our sundials and Sun compasses gives us celestial north. These can be significantly different from each other, depending on your location on Earth. (See suggestion 1 under Cross-Curricular Connections at the end of this chapter.)

HAVE A NICE DAY

The day (as defined by how long a planet takes to rotate relative to the distant stars) is different on other worlds. Jupiter has the shortest day, coming in at less than 10 hours. The day on our neighbor planet Venus is 243 of our Earth days, and to make things worse, Venus rotates backward from the direction of the other planets in our solar system. Our other neighbor, Mars, has a day that is 24 hours and 40 minutes long, tantalizingly similar to, yet annoyingly different from, our own. The engineers who control and drive the rovers on Mars have to keep Mars time, meaning their lives get out of synchronization with the lives of their families by 40 minutes per day!

Back on Earth, the Sun appears highest in the sky at local noon, although it is not generally at the overhead point (or zenith) of the sky. How high the Sun is above the horizon at any given time of day depends on your location (latitude) on Earth. More information and experiences related to how the Sun's height above the horizon varies throughout the year and based on one's latitude are covered in Chapter 2.

FIGURE 1.5

The complex nature of the Earth's magnetic field causes the location of magnetic north to migrate over time.

Magnetic north has moved from northern Canada to much closer to the North Pole over the past 184 years. Also shown are the predicted locations of the magnetic pole going back to before 1600, based on indirect measurements of the Earth's magnetic field.

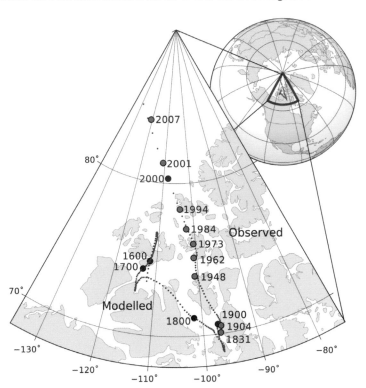

KEEPING TIME AND THE SUN

Since the Earth is a sphere and rotates, different parts of the Earth (or *longitudes*—which are lines going from the North Pole to the South Pole) will directly face the Sun at different times. For much of human history, each locality kept its own time, determined by the Sun's location in the sky. But in the 19th century, as the railroads began to regularly move populations from one place to another, it became clear that timekeeping on our planet needed to be better organized. The engineer of each train kept time based on the local time of its departure city. Nothing was coordinated, and different trains were on different schedules, which lead to unnecessary accidents. If only for the safety of travelers, countries needed to find a way to regularize time measurement. After several international meetings, the world's countries identified a series of standardized time zones, each taking up a strip 1/24th of the circumference of the Earth (or 15° of longitude).

Since it was British clock making that enabled the measurement of longitude at sea, England became the logical place for the *prime meridian*—the line of longitude from which time around the world would be measured. Time at the Royal Observatory, Greenwich, near London is now called Universal Time, and each time zone measures its time relative to that Universal Time.

The use of standard time zones is great for trains, planes, and intercity communication, but it does present problems for astronomers and classrooms wanting to study the Sun and sundials. *Noon* defined by standard time is not the same as local noon, which is defined as when the Sun crosses the north–south line in the sky in each location. If you live near the eastern edge of a time zone, local noon will occur before noon defined by standard time. Similarly, local noon will occur after noon as given by standard time at a location near the western edge of the time zone.

The concept of idealized time zones—each 15° of longitude wide—makes for good timekeeping, but it does not always match well with geographic features and cultural or national needs. The country of India spans two time zones, but in the interest of having a single time across the entire country, India "split the difference" between the time zones passing through the country. Thus, all of India is 5 hours and 30 minutes later than Universal Time. This means that when it is noon in London, it is 5:30 p.m. in India. See the Cross-Curricular Connections at the end of the chapter for suggestions for your students to identify other places where geography or culture has trumped time zone regularity.

An artistic, not-to-scale depiction of the Sun, the Moon, and the Earth

EXPERIENCE 1.1

Cast Away: What Do We Think We Know?

Overall Concept

All chapters in this book begin with an opportunity for students to reflect on what they already know about the subjects in the chapter. In this chapter, you gain an understanding of what your students already know and whether they have any preconceptions that need to be dealt with regarding the daily motion of the Sun in the sky and how it relates to our daily activities. Among the common preconceptions are that the Sun always rises in the exact direction of east and sets in the exact direction of west and that the Sun is directly overhead at noon every day of the year.

Objectives

Students will

1. reflect on and write in their astronomy lab notebooks ways that Chuck could measure the passage of time in *Cast Away*;

2. reflect on and write in their astronomy lab notebooks what they think they know regarding

   a. how the Sun's position in the sky changes throughout the day and what causes that change in position,

   b. how different the motion of the Sun in the sky is for people at other locations (both elsewhere in the United States and in places outside of North America), and

   c. what humans have done to deal with the time differences that exist between different locations; and

3. share their ideas with other students.

MATERIALS

For the class:

- Whiteboard, blackboard, or poster paper where student preconceptions can be recorded and kept visible for the duration of the time spent researching the Sun's daily motion

- (*Optional*) DVD player and a copy of *Cast Away*

One per student:

- Astronomy lab notebook

Advance Preparation

Identify space where a list of student preconceptions can be located for the length of time the class is researching the Sun's daily motion. If you show a portion of the movie *Cast Away*, have the movie advanced in your DVD player to the section you wish to show (see suggested portion under Procedure).

Procedure

1. Show parts of *Cast Away* or tell students the story of the key sections of the movie that lead to Chuck wanting to record the passage of time. (The Wikipedia entry for the movie has a plot summary.)

   Teacher note: If you show the movie, we recommend two possible sections. Watching the movie from approximately minute 31 to minute 37 nicely sets the stage for the students' conversation regarding how to measure the passing of time without a clock. A 45-minute segment (minute 31 to minute 75) shows several excellent examples in which Chuck uses scientific, engineering, and mathematical thinking to solve a number of problems he encounters. This section also provides some hints regarding timekeeping methods (e.g., lunar phases). Please note that there are some vivid images that the student will see, especially from minute 46 to minute 52, when Chuck discovers a dead colleague. You will want to review the movie, which is rated PG-13, before deciding how much to show.

2. Point out that Chuck needed to find a way to record the passage of time without a clock or calendar. Use the think-pair-share process discussed in the Introduction (p. xix) and have the students identify all the ways they can think of that Chuck could tell time, encouraging them to consider both short and long periods (e.g., seconds, minutes, hours, days, years). *Examples of possible answers include his own pulse, tides, weather patterns, Sun motion in the sky (both daily movement and annual changes), lunar phases and motion, hair and fingernail growth, and constellation positions in the night.*

3. Student suggestions will typically include the motion of the Sun throughout the day, so use this to shift to the "What Do We Think We Know?" portion of the activity. Have the students label the top of a page in their astronomy lab notebooks with "What I Think I Know." Right below that, have them write the first question you want them to consider and discuss: "How does the Sun's position in the sky change throughout the day, and what causes that change in position?" At the top of the next page, have them write the next question: "How different is the motion of the Sun in the sky for people at other locations, both elsewhere in the United States and in places outside of North America?" Finally, on the third page, have them write the last question: "You have a cousin who lives in England, and you want to talk with her. What have people done about time in different countries to make it so that you can arrange a specific time to talk on the phone?"

4. Use the think-pair-share process to (1) have students individually write answers to the questions in their astronomy lab notebooks, (2) have them discuss their answers with other students in small groups and add more detail to their astronomy lab notebooks as desired, and finally (3) write a list of what they think they know in a location in the classroom where the information can be kept for future reference.

5. Finish the experience by thanking the students for sharing what they think they know and emphasizing that keeping the answers up will help the class see what new things they learn during the study of the Sun's daily motion.

EXPERIENCE 1.2

Your Personal Pocket Sun Clock

Overall Concept

Our most basic measure of time is determined by the apparent motion of the Sun across the sky throughout the day. In this activity, students build a simple Sun clock (Figure 1.6) and explore its characteristics to spark their interest in understanding how we measure time and how this relates to the Sun's daily motion across the sky.

Objectives

Students will

1. construct a personal pocket Sun clock,

2. make observations using the pocket Sun clock to identify changes that occur with the passing of time, and

3. use their observations to begin discussing how the pocket Sun clock works and how this relates to the motion of the Sun and our concept of time.

Advance Preparation

1. Make a copy of the pocket Sun clock pattern for each student. Use the appropriate clock pattern for your location (Figure 1.7). If possible, copy the pattern on heavy paper, tagboard, or cardstock, which will eliminate the need for glue. If this is not possible, copy the pattern on regular paper and have the students glue the pattern onto heavier-weight paper. Old file folders or large index cards are good options.

MATERIALS

For the class:

- Glue

- Sticky tape

- Chalk or pencil

- (*Optional*) Compasses

One per student:

- Pocket Sun clock pattern (for your location; see Figure 1.7, p. 20)

- Card stock (e.g., file folder, index card) slightly larger than the Sun clock pattern

- String, approximately 20 cm (7 in.) long

- Scissors

- Astronomy lab notebook

2. Scout out a location that will be in the Sun when you plan to do the experience with your students (bringing a compass with you). It should be a place where they can make chalk marks on a flat surface. While you are there, get oriented so you will be able to show students the compass directions in that location (north, south, east, west).

Procedure

1. Distribute copies of the Sun clock pattern to students. Have students cut out the rectangular pattern (after gluing it to heavy paper if necessary).

2. Students then cut, as accurately as possible, the short notches at each end, as indicated on the Sun clock pattern. They should fold the clock along the dotted line on the pattern, making sure the print is on the inside.

3. Have students place one end of the string through one of the notches on the Sun clock and tape the string's end to the back of the clock.

4. Have students stretch the other end of the string through the notch at the opposite end of the Sun clock. Adjust the string so it is tight when the two panels of the clock are at a 90° angle. Then tape the string's end to the back of the Sun clock. A properly assembled Sun clock is shown in Figure 1.6.

FIGURE 1.6

A diagram of an assembled Sun clock

5. Have students work in small groups and record in their astronomy lab notebooks (followed by a whole-class discussion) what they need to do to make their clocks work (besides finding a sunny location, this might include discussion of what orientation of the clock will work best).

   Teacher note: This is an *engage* experience, so the goal is not for the student to have the "right" answer at

the end of the activity but to spark their interest to learn more about Sun clocks and the motion of the Sun. Full understanding of how the pocket Sun clock works will emerge out of the *explore* and *explain* experiences.

6.   Before going outside to use the Sun clocks, check the time on a clock or watch. If it is daylight saving time (DST), subtract one hour to get standard time. The Sun knows nothing about DST since DST is a human invention to give us more sunlight in the evening hours.

7.   Once students have the correct standard time, take them outside to a sunny location where there is a flat surface. Be sure the location will be in the sunlight for at least the next 45 minutes. Have them place their Sun clocks on a flat surface, with the string of the Sun clock taut. Have students rotate their clocks until the shadow of the string reads the correct time. Ask them to find as many orientations as possible in which their clocks to read the correct time (they can usually find at least two).

8.   Have students use a pencil or piece of chalk to draw a box around the base of the clocks so that they can tell exactly how the Sun clocks are oriented. They should put their initials inside the boxes so they can find where they had previously placed the clock when they make the next observation. Chalk works best on concrete or asphalt.

9.   Return to the classroom and ask students how they think they will need to move their Sun clocks so that the clocks tell the correct time in 15 to 45 minutes. Will all orientations work? Will they need to change their clock's orientation? How much will they need to move them? Have the students record their answers in their astronomy lab notebooks, along with the reasoning behind their predictions.

10.   After 15 to 45 minutes, have students place their Sun clocks back in the spots marked earlier and determine what must be done to read the correct time.

Teacher note: Only one orientation will work, and clocks need to be placed in the exact same position as before. The clock face points to the south with the 12 o'clock line indicating the directions north and south, which is related to how the Sun

appears to move across the sky as the Earth rotates. This is not information that needs to be explained now, but it will be discovered in the *explore* and *explain* experiences.

11. Have students work in small groups to discuss how to correctly orient the clocks. Did all orientations work? Did the direction the clocks faced have to change? Is there anything special about the direction the string faces when reading the correct time? Have the students record their conclusions in their astronomy lab notebooks.

12. In their astronomy lab notebooks, have students write instructions for using the pocket Sun clock that can be understood by anyone seeing the clock for the first time—such as a parent when the student takes a pocket Sun clock home.

13. (*Optional*) Have the groups share their instructions for using the Sun clocks and write the directions on the board as they provide each step.

Ideally, the personal pocket Sun clock experience provides the incentive for students to want to learn more about the Sun's motion in the sky and to consider what they already know about the Sun's motion, which is the focus of the next *engage* experience.

FIGURE 1.7

Templates for three different Sun clocks

Each for a different latitude in the United States*

CLOCK 3

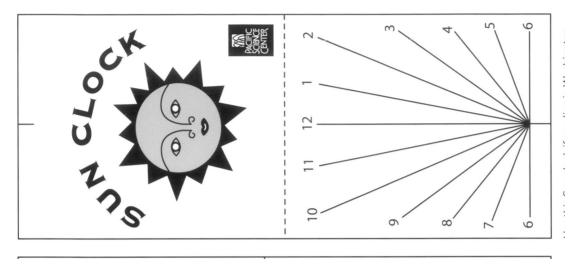

Use this Sun clock if you live in Washington, Oregon, Idaho, Montana, North Dakota, South Dakota, northern Wyoming, Minnesota, Wisconsin, Michigan, upper New York, Vermont, New Hampshire, Maine, or southern Canada.

CLOCK 2

Use this Sun clock if you live in Northern California, northern Nevada, Utah, Colorado, southern Wyoming, Nebraska, Kansas, Iowa, Missouri, Illinois, Indiana, Ohio, Kentucky, Virginia, West Virginia, DC, Maryland, Delaware, New Jersey, Pennsylvania, lower New York, Massachusetts, Connecticut, or Rhode Island.

CLOCK 1

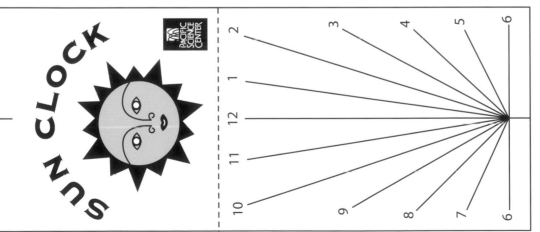

Use this Sun clock if you live in Southern California, southern Nevada, Arizona, New Mexico, Oklahoma, Texas, Arkansas, Louisiana, Tennessee, Mississippi, Alabama, Georgia, Florida, North Carolina, or South Carolina.

*These Sun clocks do not work in Alaska or Hawaii.

Three-Dimensional Learning Exposed

Three-dimensional learning refers to the importance of teaching the disciplinary core ideas, science practices, and crosscutting concepts as an integrated whole if the learning in the classroom is to be fully aligned with the *Next Generation Science Standards*. The experiences in this book allow for making explicit connections between these three dimensions.

In this chapter, students will most easily recognize the learning they do related to the disciplinary core ideas, specifically that shadows made by the Sun can be observed, described, predicted, and explained with models to reveal patterns of the Sun's apparent motion.

The various experiences in the chapter also integrate the science practices and crosscutting concepts as students learn what we know about the apparent motion of the Sun throughout the day. Especially in the *explore* and *explain* experiences, students are asked to engage in the following key science practices:

- Analyze and interpret data from their toilet-plunger sundial

- Use a model Earth and Sun (a lightbulb and their heads) to describe the relationship between the Earth and Sun and the data from their toilet-plunger sundials

- Engage in argument based on evidence as they compare the data from their toilet-plunger sundials and the observations from the mini sundial setup in "Noontime Around the World"

These experiences also allow the teacher to identify crosscutting concepts embedded in the learning. The key crosscutting concepts include the following:

3D
LE

- Patterns observed in the experiences can identify cause-and-effect relationships, as seen in the position of the Sun in the sky and the length of the sundial gnomon.

- Science assumes that objects and events in natural systems occur in consistent patterns that are understandable through measurement and observation, as seen in the position of the Sun in the sky and the length of the sundial gnomon.

- System models provide an opportunity for understanding and testing ideas, as provided in the head and lightbulb model of the Earth–Sun system, or the globe and mini sundial arrangement in "Noontime Around the World."

Although the science practices and crosscutting concepts are implicitly introduced by what the students do in the experiences, we suggest you take time during the experiences to emphasize the value of the science practices and crosscutting concepts in all areas of science.

EXPERIENCE 1.3

Shadow and Sun Tracking

Overall Concept

Now that you have sparked student interest in learning more about the motion of the Sun in the sky, you can use this activity to provide students with the opportunity to research the Sun's apparent motion across the sky by studying the shadow cast by the Sun and observing the location of the Sun in the sky. Students use a simple gnomon made from a small toilet plunger to produce a shadow and collect data about the changing direction and length of the shadow throughout the day. As the Sun appears to move across the sky from east to west each day, shadows change direction and length throughout the day. At noon, the gnomon's shadow is the shortest, and the shadow is pointing to celestial north (see p. 9 of the Content Background to learn how this differs from magnetic north).

> *Teacher note:* This activity consists of two elements, which are written up separately: (1) shadow tracking and (2) Sun tracking.

Objectives

Students will

1. observe the changing direction and length of a shadow over a portion of a day,

2. determine how the change observed in the shadow relates to the motion of the Sun across the sky,

3. realize how the Sun's shadow relates to directions and time on Earth, and

MATERIALS

One per group of three to four students:

- Small toilet plunger (a cheap wood-handled plunger works well; the handle should be cut to about 22 cm so that the shadow will fit on your poster-sized sheet of paper)

- Chalk and black marker

- Poster-sized sheet of blank paper

- (*Optional*) Compass

One per student:

- "Where Is the Sun Now? Observation Diagram Sheet" (p. 29)

- "Moving Shadows Discussion Questions" (p. 31)

- Astronomy lab notebook

1.3

EXPLORE

4. determine that the Sun appears to move across the sky from east to west each day and is lower in the sky in the early morning and late afternoon compared with the middle of the day.

Advance Preparation

This experience requires a period of sunshine in both the morning and the afternoon, so be aware of the weather forecast as you plan for this activity. You will need at least 60 minutes of clear skies both before and after noon. Please recall that if you are on daylight saving time, your clock time is off by one hour. Subtract one hour from what your clock shows (e.g., 1:00 p.m. is actually noon).

If possible, extend the observations to include as much of the day as possible. Have the teams do their initial setup in the morning and record the shadow every half hour until 11:00 a.m. (standard time). They then need to record the shadow every 5 to 10 minutes, as noted in the Procedure. The more data collected throughout the day, the better your students will understand the pattern of the Sun's motion.

If you teach multiple classes throughout the day, you will need to be creative to complete this experience. If you have a number of classes doing this activity, have them share their observations of the gnomon's shadow, with classes early in the day providing morning observations and classes later in the day adding noontime and afternoon observations on the same poster paper. If you teach only one or two classes during the day, identify students who are willing to go out during lunch or other free time to make the observations for the entire class. Encourage each student to go out regularly throughout the day to make observations of the Sun's location on their "Where is the Sun Now? Observation Diagram."

Just before starting the experience, tell students they will be collecting and analyzing data about the Sun's location in the sky and the shadow that the Sun produces of the gnomon on a sundial. They will use this data to deduce what they can learn about the motion of the Sun and Earth and how we use this relationship in our daily lives. Explain whether they will be gathering data throughout the day or cooperating with other classes to gather data at times they are taking a different subject.

Safety note: Emphasize to your students that they will be making an indirect observation of the Sun. They should never look directly at the Sun because even a brief viewing can damage the retinas of their eyes.

Procedure for the Shadow Tracking Element

1. Divide the students into teams of three to four students. Ideally, observations are made throughout the day, but the minimum time needed to do this activity is about 120 minutes, with outdoor setup starting at 10:30 a.m. Observations should start by 11:00 a.m. standard time. You will want your observations to span from before until after local noon.

   Teacher note: Local noon is defined in the Content Background section of this chapter on page 11. Local noon can vary significantly from standard time noon because of your location within your time zone. You will find it useful to practice this activity yourself at home before doing it with the entire class. This will also give you a set of data taken in a different location and on a separate date to share with students.

2. Before going outside, make sure each team has a large piece of paper and a toilet-plunger gnomon. Be sure each plunger handle sticks straight up out of the base. Give students a chance to get over the hilarity of using something associated with a toilet to do astronomy. Tell them it just goes to show that the tools of science can be found everywhere.

3. Have the teams spread out in a large clear area (a cement or asphalt area of the school works well) that will be in the sunlight for the duration of the observations. Students should place the gnomon near the middle of the long edge of the paper so that the shadow of the top end of the plunger falls on the paper (see Figure 1.8). Students need to draw a line around the plunger base so it can be replaced in the right location if the plunger gets moved.

4. Students also need to use the chalk to draw an outline on the concrete or asphalt around the entire sheet of paper. An *X* in one corner of the paper and the outline on the concrete or asphalt will allow the students to bring the paper back to the same place and orientation for later observations if needed. They should also put their names on the paper so that observations can be saved for use in experiences in Chapter 2, when students make

observations to determine how the Sun's apparent motion changes over the course of a year.

5. Once the setup is ready, the students mark the very end of the gnomon's shadow on the paper. They write the time of the observation next to the shadow's end each time they make an observation (see Figure 1.8).

6. Students continue to mark the end of the shadow at 5- to 10-minute intervals over a period of at least 60 minutes that span over noon. Continue the observations until at least 12:30 p.m. standard time and longer if possible. The more observations made, the more obvious the change in the shadow's length and direction will be.

Teacher note: When viewed from behind the gnomon, the shadow will move from left to right and will change length. If you make the observations before and after solar noon, the shadow will shorten and then lengthen. Remember that during DST, your clock time is off by one hour.

FIGURE 1.8

A diagram showing equipment setup and a student marking the end of the shadow

NATIONAL SCIENCE TEACHERS ASSOCIATION

7. Have the teams bring all equipment back to the classroom after the last observations are made. Have students draw rough sketches of their observations in their notebooks, making sure they capture the changing direction and length of the shadow. Remind them that they need to keep their observations safe in their notebooks so they can be used in the Chapter 2 experiences. You should collect the large sheets of paper for future use during Chapter 2 activities.

Procedure for the Sun Tracking Element

1. Students paste the "Where is the Sun Now? Observation Diagram Sheet" into their notebooks.

2. Have students follow the instructions on the "Where is the Sun Now? Observation Diagram Sheet" to observe where the Sun is in the sky throughout the day. If possible, students should make the observations from the same general location where they placed their toilet plunger sundials, but it is not absolutely necessary (see Teacher Note below).

3. When you go outside with the students to make the first observation, orient them to the directions east, south, west, and north. If you do not want to just tell them the directions, have them use a compass to roughly get the correct direction. Have the students draw in key stationary objects on the diagram (e.g., trees, buildings, flagpoles) and then make the first observation of the Sun's position relative to the stationary objects.

   Teacher note: We say "roughly get the correct direction" because, depending on your latitude, magnetic north (measured by a compass) can be significantly different than astronomical north (as measured by the apparent motion of the Sun across the sky). This activity requires a reasonably clear view to the south, so you will want to do the activity yourself for a few days to make sure the location you choose is a good place to observe the changing location of the Sun in the sky. Also, be sure to remind students that it is not safe to look directly at the Sun for an extended length of time. They should just briefly glance so that they can plot the Sun's position on the diagram.

4.   Have students continue to go out each hour until the end of the day to mark the position of the Sun on the "Where is the Sun Now? Observation Diagram."

Procedure After Observations Are Collected

1.   Hand out copies of the "Moving Shadows Discussion Questions" worksheet. In their teams, students should write answers on the worksheet and discuss them in their groups. Have them tape the worksheet in their astronomy lab notebooks. Then have a whole-class discussion and record the consensus to the answers on the board for later reference. As the students discuss their answers to the questions, be sure to ask them to back up their conclusion with evidence based on their observations.

2.   At this time, it is not necessary to resolve any differences of opinion. Just let the students argue their points of view based on the evidence they collected. Resolution of differences should occur as you work with students during the *explain* experiences in the next section.

3.   When you are ready to move on to the *explain* experiences, a discussion of the two transition questions at the end of the worksheet are useful lead-ins to the first *explain* experience.

Where Is the Sun Now?

OBSERVATION DIAGRAM SHEET

Record your observations of the Sun for your location in the diagram below. This wide-angled perspective shows your view of the horizon looking south from where you placed your toilet-plunger gnomon or from other nearby locations where your view is not too obstructed. Your teacher will help you establish where the cardinal directions (the directions of the compass) are in your location. Stand so east is to your left and west is to your right.

1. On the diagram, fill in some stationary objects that you see in your view of the horizon. These may be buildings, trees, telephone poles, or other objects of your choosing. Start by drawing an object that is close to south in your diagram. Then draw an object in the east, followed by one in the west. Next, fill in few a more objects in between those that you already have. One tree is shown to demonstrate the approximate size of things for your drawing.

2. Each time you make an observation, note the location of the Sun in the sky in relation to the objects in your drawing. Do not look directly at the Sun because looking at it can hurt your eyes. Just glance at its approximate location in relation to objects in your drawing. Label each Sun location with the time of day you made the observation. If you are doing this activity during daylight saving time, remember to record the time in standard time by subtracting one hour from the time on your clock.

3. Wait until the next observation time and record the Sun's location in your diagram, again labeling the time. Continue making as many observations as you can throughout the day.

4. Describe in your astronomy lab notebook the motion of the Sun across the sky throughout the day.

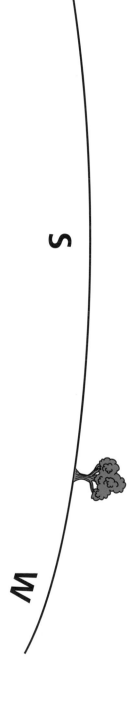

E

S

W

Moving Shadows Discussion Questions
TEACHER VERSION*

1. Describe what happened to the gnomon's shadow over the course of the time you observed it.
   The shadow moved across the paper. It got shorter, and then it got longer.

2. What time was the shadow the longest? What time was it the shortest?
   The shadow was longest at the earliest and latest observation times. It was shortest around local noon. (Remember, for DST, this would be around 1:00 p.m.)

3. What did the Sun do in the sky from the start to the end of your observations? Describe its movement.
   For example, the Sun moved from over the big tree in the school yard to over the school library. It also got higher in the sky and then lower. (Students may have difficulty seeing the movement if they observe for only a short period of time.)

4. Which way did the gnomon's shadow move when looking south from behind your gnomon? Did it move from right to left or from left to right? Which way did the Sun seem to move?
   The shadow moved right to left, and the Sun seemed to move across the sky from left to right.

5. Describe the movement of the shadow and Sun in terms of the cardinal directions (north, south, east, and west).
   The Sun seems to move from east to west. The gnomon's shadow moved from west to east.

6. What is the relationship between the shortest shadow you observed and the position of the Sun in the sky? How does this relate to the compass directions north, south, east and west?
   The shortest shadow occurred when the Sun was highest in the sky and was directly south of our location. (This is local noon.)

Thought questions for consideration as a transition to the *explain* experiences:

7. You may have heard that the Sun doesn't really move across the sky; rather, the Earth is spinning and making the Sun appear to move across the sky. From your viewpoint, is the Earth spinning toward the east or toward the west?
   Earth is spinning toward the east. (It is not critical to reach a final conclusion at this time because students will explore and discover the answer to this question during the coming experiences.)

8. If you were in outer space above the North Pole looking down on the Earth, would you see Earth spinning clockwise or counterclockwise?
   The Earth is spinning counterclockwise. (Again, it is not critical to reach a final conclusion at this time because students will explore and discover the answer to this question during the coming experiences.)

Contains key information that could be included in responses. For a copy of the student handout with more space for answers, please visit www.nsta.org/solarscience.

Moving Shadows Discussion Questions

1. Describe what happened to the gnomon's shadow over the course of the time you observed it.

2. What time was the shadow the longest? What time was it the shortest?

3. What did the Sun do in the sky from the start to the end of your observations? Describe its movement.

4. Which way did the gnomon's shadow move when looking south from behind your gnomon? Did it move from right to left or from left to right? Which way did the Sun seem to move?

5. Describe the movement of the shadow and Sun in terms of the cardinal directions (north, south, east, and west).

6. What is the relationship between the shortest shadow you observed and the position of the Sun in the sky? How does this relate to the compass directions north, south, east and west?

Thought questions for consideration as a transition to the *explain* experiences:

7. You may have heard that the Sun doesn't really move across the sky; rather, the Earth is spinning and making the Sun appear to move across the sky. From your viewpoint, is the Earth spinning toward the east or toward the west?

8. If you were in outer space above the North Pole looking down on the Earth, would you see Earth spinning clockwise or counterclockwise?

EXPERIENCE 1.4

Modeling the Sun–Earth Relationship

Overall Concept

Students have made observations and data related to the apparent motion of the Sun across the sky throughout the day. This activity now provides experience with a physical model that uses their heads as the "Earth" and a bare lightbulb as the "Sun." This helps the students develop a mental model to explain the motion of the Sun and shadows made by the sundial. It also helps them understand that the Earth's rotation on its axis causes the Sun to appear to move across the sky. It demonstrates and explains the concepts of noon, midnight, sunrise, and sunset and sets up the transition to the next experience. The model this experience introduces will be used again in Chapter 4 to show the relationships among the Earth, Sun, and Moon and in understanding what causes eclipses.

Objectives

Students will

1. participate in a modeling activity of the Sun–Earth system and

2. be able to describe why the Sun seems to move across the sky in a recognizable pattern every day.

Advance Preparation

Find a room that can be made fairly dark for this experience.

MATERIALS

For the class:

- Lightbulb on a stand or clamp (or a lamp with the shade removed)

- Extension cord

- Dark room (large enough for students to spread out and see the light)

- A globe of the Earth

- Sticky notes (two per student)

One per student:

- Astronomy lab notebook

Procedure

1. Explain that in science we often have to model phenomena because they are too big or complex to study in the real world. The relationship between the Sun and Earth is a good example of when modeling can help us understand what's occurring. Students can be part of this model. Their heads represent the Earth and their noses their location on it. The top of each student's head is the North Pole while the base (the bottom of the chin) is the South Pole. The Sun, represented by the lightbulb, is placed high in the front of the room.

2. Turn on the lightbulb and turn off the room lights. Students should stand and face the lightbulb. Make sure all students can see the "Sun" and that their faces are not hidden in shadows. Explain that their noses represent their hometown. It is often helpful to have them thinking about reshaping their nose into a local iconic building or feature (e.g., in Seattle this could be the Seattle Space Needle or Mount Rainer).

3. Ask students what time it would be at their location if their noses were directly facing the Sun. Give them some time to think about this and discuss it with their neighbors.

4. Students should recognize from the previous lesson that the Sun will be visible and the highest in the sky at about noon.

5. Once students understand the idea of "noses-at-noon," have them stand so that their location (the nose) is at midnight. After discussion with other students, they should ultimately all turn to face directly away from the light.

6. When students turned from noon to midnight, some students probably turned clockwise while others turned counterclockwise. Point this out and ask students which way Earth actually turns on its axis. You should have a globe available to help students with modeling this idea. From the *explore* experience, students may remember the Sun moved from east to west during the time they observed it. On the globe, have students locate east of where they are and west of where they are. The East Coast and West Coast of the United States work well for students in the

continental United States. Discuss the fact that people in the east will see the Sun rise before those in the west. Show students how this can happen using the globe and lightbulb. See if you can have them come up with the conclusion that the Earth must turn from west to east (in an easterly direction).

Teacher note: Looking down from the North Pole, the Earth spins counterclockwise. Therefore, the student's Earth (his or her head) should turn toward the left, with his or her right shoulder moving forward toward the left. See Figure 1.9.

7. Have students turn counterclockwise so they go through a complete 24-hour period, starting with noses-at-noon. Make sure all students turn the proper direction.

8. Ask students to turn so their noses are at about 6:00 p.m. They should all turn left so their noses are at about a 90° angle to the lightbulb, with the right sides of their faces illuminated. Ask them where the Sun is in the sky for them. It is just about to set and is seen very far down on the western horizon.

Teacher note: For this model, it is assumed that there are 12 hours of daylight and 12 hours of night. Students may bring this up, particularly if you live in an area where there is a dramatic difference in the amount of daylight throughout the year. Explain to them that this is a difficulty with using models. Sometimes we simplify things to understand an idea, and the model may not be fully accurate. The changing amounts of daylight throughout the year will be addressed in later experiences.

9. Have students locate various times of the day by turning around. Make sure all students are turning in the correct direction at all times and they are facing approximately the same direction for a given time. This is a good time to introduce the concept of horizon blinders (see Figure 1.9). If students hold their hands flat, palms front, one on each side of their heads, they can use these blinders to help them see the Sun disappear and reappear as they turn through the 24-hour day. If students are having trouble keeping track of west and east on their model Earth (their heads), you may find it useful to put sticky notes labeled *east* and *west* on each hand. East goes on the left hand and west goes on the right hand.

FIGURE 1.9

Horizon blinders

Student turns this way

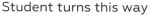

10. Reinforce the observations students made in the *explore* experiences by focusing their attention on the changing position of the Sun in the model. Remind them that Earth takes about 24 hours to spin one full revolution on its axis. The Sun moves the same angular distance across the sky during any given time period (about 15° each hour). The Sun is highest in the sky at solar noon (this may differ from watch time because of daylight saving time and also because of your location within your time zone).

11. To solidify and evaluate student understanding, have students write a short essay in their astronomy lab notebooks explaining what they have learned in a letter to one of their grandparents. Alternatively, simply have them answer the following questions:

    a. How long does it take Earth to spin once around on its axis?

    b. When is the Sun at its highest point in the sky at your particular location?

    c. Where is the Sun located at different times of the day (e.g., 8:00 a.m., noon, 3:00 p.m., and midnight)?

    d. How would you describe the direction of the spin of the Earth on its axis?

EXPERIENCE 1.5

Noontime Around the World

Overall Concept

Small groups of students are given a globe, toothpicks, and balls of clay or putty. Students will use a toothpick and clay to make a gnomon for their globe and mark their location. By making the globe rotate, they can see the same change in shadow length and direction that they saw in their own observations outside. More gnomons can be added to the globe to see how the shadows at a certain time vary for other locations. This also allows students to explore the difference in time for a number of locations on the Earth. A full page of gnomon recording circle templates is available at *www.nsta.org/solarscience*.

Objectives

Students will

1. build a model sundial on a model of the Sun–Earth system;

2. understand how the rotation of the Earth explains the daily motion of the Sun across our sky and the shadows it produces; and

3. understand that when it is noon in their location, most other places on the Earth are experiencing a different time of day.

Advance Preparation

1. Make sure the room can be made relatively dark—although it does not need to be as dark as for Experience 1.4, "Modeling the Sun–Earth Relationship."

MATERIALS

One per group of three to four students:

- Clamp lamp with directional shades that cause the light to shine in one direction

- Globe

Several per group:

- Small balls of clay or picture-hanging putty

- Ends of toothpicks (cut to 8 mm long)

- Gnomon shadow recording circles (see template in Figure 1.10)

One per student:

- "Noontime Around the World Discussion Questions" (p. 42)

- Astronomy lab notebook

2. Be sure students have enough room and table space to set up their Sun–Earth systems so their model Earths and Suns are at least four feet apart.

Procedure

1. Have students build a model sundial from a ball of putty that is about the size of small pea, a gnomon shadow recording circle (Figure 1.10), and a toothpick, as shown in Figure 1.11.

2. Have students set up their globe and lamp with the lamp's shade directing the Sun's light on the Earth, and with the lamp at the same height above the table as the Earth's equator (see Figure 1.12, p. 38). The axis of rotation of the globe should be pointed away from the direction of the Sun (i.e., with the North Pole tipped away from the direction of the Sun), which represents the Earth's orientation relative to the Sun on the winter solstice (the first day of winter in

FIGURE 1.10

Gnomon shadow recording circle template

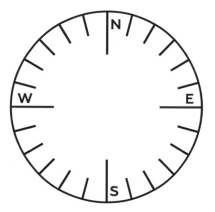

FIGURE 1.11

Diagram showing how to build a model sundial

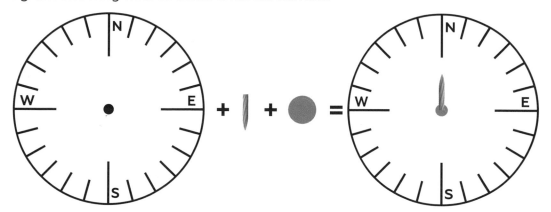

Cut out enough "Gnomon Shadow Recording Circles" so that each group can build three sundials. Put a small hole in the center so that the toothpick base can stick through the hole. The toothpick sticks up through the hole with the putty underneath, which will hold it in place and attach it to the globe.

FIGURE 1.12

A model Sun and Earth setup

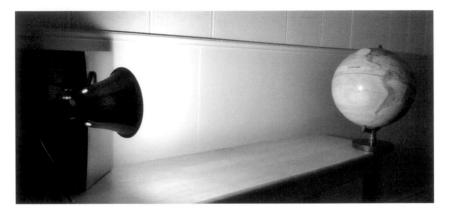

Note that the model Sun is pointed directly at the globe's equator.

the Northern Hemisphere, which occurs around December 21).

3. Have them attach the sundial to the globe at their location. Be sure the north–south line on the dial is lined up to point to the North and South Poles of the globe, and the toothpick gnomon sticks straight out from the surface of the globe (Figure 1.13).

4. Put the following questions on the board or project them onto a screen or wall. Have students explore with their model Sun–Earth system to answer the questions in their astronomy lab notebooks (be sure they always turn the globe in a counterclockwise direction when looking down on the North Pole):

   a. Turn your globe so that it is sunrise for your model sundial. What is the direction of the model Sun from your model sundial? *Toward the east.*

FIGURE 1.13

A model sundial at your location

Note that the north–south line is lined up with north and south on the globe.

b. What is the direction of the model Sun from your model sundial when it is sunset at your location? *Toward the west.*

c. Where is the Sun relative to the sundial when it is noon? *Toward the south and higher in the sky than at sunrise or sunset.*

d. Which part of the United States experiences sunset first? Which part experiences it last? *For the continental United States, the East Coast experiences sunset first and the West Coast last. If students ask about sunrise and sunset in Hawaii and Alaska, encourage them to use the model to answer the question.*

5. These four questions give you time to observe the groups to see how much they understand and who needs extra help with the model.

   Teacher note: Depending on what globe you use, the support arm holding the globe in place may not allow the gnomon to clear the support arm as the globe is turned. Students will need to adjust the gnomon to stick straight out after the sundial passes under the support arm. Also, this experience can be a challenge for students because of the spatial reasoning required. It will help if they think of themselves as miniature people standing next to the sundial as they work to answer these questions. Finally, students may have a hard time knowing exactly when sunrise and sunset occur on their globes. The best indicator of when a location on the globe has moved from daytime to nighttime is when the students can no longer see a shadow from the gnomon. Sunrise is when the students can first see the gnomon's shadow.

6. Have a whole-class discussion of what the students discovered, reinforcing that the Sun rises along the eastern horizon and sets along the western horizon. Additionally, the sundial is facing generally toward the direction of the Sun at noon with the shadow directly along the north–south line.

7. This is a good time to discuss the idea of time zones with the class. Because at any given time the Sun is at different positions in the sky for different parts of our country and of the world, we divide the globe into time zones. Ask the class if they know what time zone the school is located in. Then ask them to think about whether their family has relatives or friends who live in different time zones. What do they need to keep in mind when they call such relatives or friends?

8. Have the students work in their groups to complete the "Noontime Around the World Discussion Questions" worksheet. They may find it useful to use the additional recording circles, toothpicks, and balls of clay to make mini sundials to attach to other locations on the globe. Have a whole-class discussion after they complete their group discussion and have them attach the worksheet in their astronomy lab notebooks.

Noontime Around the World
Discussion Questions
TEACHER VERSION*

1. What happens to the shadow on your sundial as you rotate the globe to take your location from sunrise to noon to sunset? How does this compare with what you observed with your toilet-plunger sundial?

    At sunrise, the shadow is relatively long and pointing generally toward the west from the base of the gnomon. At noon, the shadow is shorter and points north in the direction of the North Pole. Near sunset, the shadow is longer again and points generally toward the east. This is much like what students observed outside using the toilet plunger as a gnomon.

2. What location in another country on your globe has noon at the exact same time as you? What is your evidence that this is true?

    Students should identify a location on the same longitude line as their home location. Students can build a mini sundial to attach at another location to test their ideas. This will show that the gnomon's shadow falls along the north–south line when it is noon anywhere along the same longitude line as their home location.

3. How does the sunrise and sunset time for that location in another country differ from the time at your location? *The answers will depend on what location they choose.*

    Since the model is set for the winter solstice, locations south of their home location will have an earlier sunrise and earlier sunset. For locations north of their home location, sunrise will be later and sunset earlier.

4. What is a location on the globe that experiences midnight when you experience noon?

    If students follow their longitude line to the other side of the Earth away from the direction of the Sun, they reach the longitude line 180° away from their home location. Any location on this longitude line will experience midnight when it is noon in the students' location.

5. What time zone in the United States experiences sunrise first? What time zone in the United States experiences sunrise last? (*Hint:* Remember that not every state in the United States is connected to where most states are!)

    The East Coast of the United States experiences sunrise first, while Hawaii and Alaska experience it last.

6. Are there places on the Earth that never experience sunrise or sunset?

    On the winter solstice, the area near the North Pole is always in the dark as you rotate the Earth, so it has no daily sunrise and sunset. The area near the South Pole is always bathed in sunlight as you rotate the Earth, so it also has no daily sunrise or sunset.

7. The Sun is unlikely to be directly overhead at your location at noon. Can you find a location where the Sun is directly overhead when it is noon for you? What is the evidence that you have found that location?

    Students should move their model sundial along their longitude line on the globe until the gnomon has no shadow. This means the Sun is directly overhead at this location.

*Contains key information that could be included in responses. For a copy of the student handout with more space for answers, please visit www.nsta.org/solarscience.

Noontime Around the World
Discussion Questions

1. What happens to the shadow on your sundial as you rotate the globe to take your location from sunrise to noon to sunset? How does this compare with what you observed with your toilet-plunger sundial?

2. What location in another country on your globe has noon at the exact same time as you? What is your evidence that this is true?

3. How does the sunrise and sunset time for that location in another country differ from the time at your location?

4. What is a location on the globe that experiences midnight when you experience noon?

5. What time zone in the United States experiences sunrise first? What time zone in the United States experiences sunrise last? (*Hint:* Remember that not every state in the United States is connected to where most states are!)

6. Are there places on the Earth that never experience sunrise or sunset?

7. The Sun is unlikely to be directly overhead at your location at noon. Can you find a location where the Sun is directly overhead when it is noon for you? What is the evidence that you have found that location?

EXPERIENCE 1.6

Pocket Sun Compass

Overall Concept

Students discovered in the first *engage* experience that they needed to orient their Sun clocks correctly to tell the time. In this activity, they realize that a Sun clock reading the correct time can be used to tell directions (north, south, east, and west). All of the directions for this activity assume the participants are in the Northern Hemisphere.

Objectives

Students will

1. discover how their pocket Sun clocks allow them to determine the cardinal directions,

2. follow a set of directions using the Sun clock as a compass, and

3. create a map using a set of directions.

Advance Preparation

Determine an appropriate area for the outdoor portion of this activity. If desired, determine some "treasure" that you can write on the back of the *X*s that the group will receive as a reward. Make a copy of the challenge handout for each student.

Safety note: Remind students that it is not safe to look directly at the Sun.

MATERIALS

One per group of two to three students:

*   Watch or phone with time functions

*   Pencil or chalk

*   Sheets of cardboard with *X*s or other marks on them, each one with a different color or design so students will know which one is theirs

*   "Treasures" (rewards) for each group, such as a science bumper sticker, astronomy-named candy (e.g., Milky Way, Starburst, Eclipse gum), or something as simple as "Congratulations" written on the back of the *X*.

One per student:

*   Pocket Sun clock (created in the first *engage* experience)

*   "Sun Compass Challenge" (p. 46)

*   Astronomy lab notebook

Procedure

1. Break students into small groups. Ask them to think about how the Sun clock could be used to find directions (north, south, east, west). Allow time for them to explore their ideas with their Sun clocks. Have them write their speculations in their astronomy lab notebooks. This extends nicely from the discussion in the earlier experiences about the need for proper orientation of the Sun clock in order for it to work correctly. Some students already may have determined that the proper orientation for the Sun clock is to have the clock face, with the hour lines, turned toward the south.

2. Write the students' speculations about how to use the Sun clock to find directions on the board, guiding the discussion toward the following conclusions:

   a. You need to know the time from a watch or clock to know that the Sun clock is properly aligned. Don't forget that you need to subtract an hour from your watch's time if it is daylight saving time.

   b. If the Sun clock is correctly aligned to read the time, it must be facing south; the rest of the directions can then be determined.

3. On a sunny day, go outside with the Sun clocks. Ask students to hold their Sun clocks level in their hands and to rotate their clocks until they show the correct time. The Sun clocks are now facing south, with the strings running in a north–south direction. Have students point in the direction you call out to them. If they are facing south, north will be behind them, east to the left, and west to the right. Try several different directions until you are confident that students know them.

4. Place your cardboard with different color or design Xs on them about 20 feet apart in an open area with little shade. You may need to place a rock on top of each X so that they do not blow away. Divide students into small work groups of two to three. Have each group stand beside an X.

5. Give each student a copy of the "Sun Compass Challenge." In order to solve it, they will have to use their Sun clocks as compasses.

   Teacher note: If students successfully follow the directions, they will end at the spot where they started. If they pick up the cardboard *X* and turn it over, they will find their treasure (if you put one behind it). More advanced students can devise their own directions for other teams to follow. Some may want to experiment with intermediate directions, such as northeast or southwest.

6. After students have worked outdoors and followed the directions to the treasure, have each student draw a map in his or her astronomy lab notebook that accurately indicates the cardinal directions and the path he or she followed. The grade level of the students can determine the accuracy of the scale of the drawing. What school landmarks might be added to their maps to provide additional references?

7. Students may wish to create their own treasure maps with directions. If some groups do this, they can challenge other groups to follow the directions and see where they end up.

Sun Compass Challenge

Challenge: A treasure can be found where these directions lead you. Use your Sun clock to follow the instructions below that tell you where you should look:

1. Take 5 steps north.
2. Take 5 steps east.
3. Take 5 steps north.
4. Take 10 steps west.

5. Take 5 steps south.
6. Take 10 steps east.
7. Take 5 steps south.
8. Take 5 steps west.

NATIONAL SCIENCE TEACHERS ASSOCIATION

EXPERIENCE 1.7

High Noon

Overall Concept

Students return to the gnomon shadow data they collected earlier in their small groups to identify when the Sun's shadow is the shortest. If necessary, they may need to collect shadow length more often around noon to identify the precise moment when the shadow is shortest. They then use simple geometry and the law of similar triangles to determine the height (in degrees) of the Sun above the horizon, especially at noon. This reinforces the idea that the Sun is not always directly overhead at noon, and will provide data that can be used with experiences in Chapter 2.

Objectives

Students will

1. measure the length of a gnomon's shadow to determine the Sun's angular height in the sky and

2. use this technique to find the Sun's height in the sky at noon.

Advance Preparation

Students need to measure the altitude of the Sun at 10:00 a.m., solar noon, and 2:00 p.m., so you will need to find a location where the Sun will be visible during the entire time. The class will need to be outside briefly for the 10:00 a.m. and 2:00 p.m. measurements and for at least 40 minutes for the solar noon measurement. The time needed to take the solar noon reading can be decreased if you are able to determine the exact time of solar noon for your location using an almanac, newspaper, or ephemeris. A solar noon calculator is also available at *www.esrl.noaa.gov/gmd/grad/solcalc*.

MATERIALS

One per group:

- Large blank sheet of construction or chart paper (one sheet per group per measurement; at least 45 cm on one side)

- Meterstick

- Small plastic protractor

- Gnomon (toilet plunger) from *explore* Experience 1.3

One per student:

- "High Noon Data Chart" (p. 50)

- Astronomy lab notebook

1.7
ELABORATE

Procedure

1. Students measure the height of the gnomon and the length of the shadow (both in centimeters) at each designated time. The length of the gnomon is from the ground to the top of the plunger handle. The length of the shadow is from the center of the circle to the mark where the top of the gnomon's shadow is. Students record these measurements in the "High Noon Data Chart."

2. To calculate the altitude (degrees above the horizon) of the Sun, students transfer their data onto their large construction or chart papers (see Figure 1.14). It is best to use one piece of paper for each observation made, although the use of different colored pens or pencils can allow for multiple observations on one sheet. A horizontal line the length of the shadow should be drawn near the bottom of the short side of the paper. Label this SG (S for shadow, G for gnomon). Using a protractor and meterstick, draw a line that is 90° (perpendicular) to SG up the long side of the paper that is the same length as the gnomon. The plunger height might vary according to brand, but should be about 25 cm (10 in.) tall. This line should start at the point labeled G. Label the other end of the line H (for height).

3. Draw a line connecting points S and H. This represents the direction of the Sun (see Figure 1.14).

4. Find the altitude of the Sun by measuring the angle HSG with a protractor. Students should record this in their data charts. One member of the group should fold their drawing and insert it into his or her astronomy lab notebook for record keeping.

5. Each student should insert the "High Noon Data Chart" into his or her astronomy lab notebook. It will be used again in the next chapter when the class examines how the Sun's motion and height above the horizon changes throughout the year.

FIGURE 1.14

Data points showing the Sun's location above the horizon on graph paper

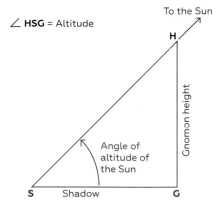

∠ **HSG** = Altitude

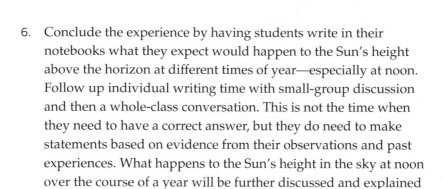

6. Conclude the experience by having students write in their notebooks what they expect would happen to the Sun's height above the horizon at different times of year—especially at noon. Follow up individual writing time with small-group discussion and then a whole-class conversation. This is not the time when they need to have a correct answer, but they do need to make statements based on evidence from their observations and past experiences. What happens to the Sun's height in the sky at noon over the course of a year will be further discussed and explained in the next chapter.

High Noon Data Chart

Use this chart to record your high noon observations.

Date	Time	Length of gnomon	Length of shadow	Sun's altitude

EXPERIENCE 1.8

Write a Picture Book for Kids

Overall Concept

Each student writes and illustrates a picture book for a younger student, explaining how the Sun moves across the sky during the day and what happens to shadows on the Earth as the day goes by. Three writing prompts are provided to encourage the inclusion of key ideas learned in the chapter.

Objectives

Students will demonstrate how well they understand

1. the apparent motion of the Sun across the sky during the day,

2. what happens to shadows of objects on the Earth throughout the day, and

3. what causes day and night.

Procedure

1. Tell students that they will write and illustrate a children's book for students two to three years younger than they are. The book must include

    a. explanations and illustrations of what causes day and night,

    b. how the Sun appears to move across the sky, and

    c. what happens to shadows of objects on the Earth throughout the day.

MATERIALS

- Blank pages for a book (e.g., pages in the student's astronomy lab notebook, sheets of paper that get stapled in the corner, or a book made from 8 ½ × 11 sheets of paper folded in half)

- Colored pencils or pens (enough so all students can have a selection), plus access to magazines, images on the web, or other resources for use in illustrating their books

The book can be either nonfiction (just the facts, nicely explained for kids) or fiction (telling a story that still gets the facts across as it follows some characters).

2. Key elements to look for in the finished books include the following:

    a. The Sun appears to move from the eastern part of the sky to the western part of the sky throughout the day.

    b. This movement of the Sun is due to the Earth rotating rather than the Sun actually moving across the sky.

    c. It is daytime when our location on the Earth rotates to face toward the Sun; nighttime occurs when our location rotates to be in the opposite direction from the Sun.

    d. Shadows point away from the direction of the Sun and are longer in the morning and afternoon when the Sun is lower in the sky.

    e. Shadows are the shortest at noon when the Sun is highest in the sky—not directly overhead—and the Sun is directly above the south direction on the horizon.

    If there is a need to give a score for the content of the book, consider one point for each of the five elements listed above.

NATIONAL SCIENCE TEACHERS ASSOCIATION

EXPERIENCE 1.9

Where Is It Night When We Have Noon?

Overall Concept

Students use the equipment from Experience 1.5, "Noontime Around the World," to demonstrate how well they can translate what they learned to a related situation. Students use the materials to answer several questions, such as where it is nighttime and midnight when it is noon at different places on the Earth, including their home location.

Objective

Students will demonstrate how well they can translate to other locations on the Earth what they learned about sundials, gnomon shadows, noon, and midnight at their home location.

> *Teacher note:* Students do not necessarily need the mini sundials used in Experience 1.5, "Noontime Around the World," to answer the questions in this experience. However, you may find it useful to have some mini sundials available if students want to use them to help answer the questions.

Advance Preparation

1. Make sure the room can be made relatively dark—although it does not need to be as dark as for Experience 1.4, "Modeling the Sun–Earth Relationship."

2. Be sure that students have the room and table space to set up their Sun–Earth systems so that the model Earth and Sun are at least four feet apart.

MATERIALS

One per group of three to four students:

- Clamp lamps with directional shades that shine the light in one direction

- Globe of the Earth

One per student:

- "Where Is It Night When We Have Noon? Question Sheet" (p. 56)

- Astronomy lab notebook

Procedure

1. Have students set up their globe and lamp with the lamp's shade focusing the Sun's light on the Earth and with the lamp at the same height above the table as the Earth's equator. This time, have the axis of rotation of the globe points toward the direction of the Sun (i.e., with the North Pole tipped toward the direction of the Sun), which represents the Earth's orientation relative to the Sun on the summer solstice (first day of summer in the Northern Hemisphere, which occurs around June 21).

2. Distribute the "Where Is It Night When We Have Noon? Question Sheet" and have students work in small groups to answer the questions. If you want to have responses from individual students, an alternative approach is to have a few Earth–Sun setups in the back of the room and ask individual students to visit the setups and respond to the questions while other activities are occurring elsewhere in the room.

Where Is It Night When We Have Noon?
Question Sheet

TEACHER VERSION*

1. When it is local noon at your location, what are two other places that are also having local noon? How do you know it is noon there?

    Students should identify two locations on the same longitude line as their home location and explain that they know it is noon because the gnomon's shadow falls along the north–south line.

2. When it is local noon at your location, what are two places that are experiencing midnight?

    Students should identify two locations on the longitude line directly on the other side of the globe from their home location.

3. Are there locations on the Earth where you would not see the Sun throughout an entire day? If so, where?

    Near the South Pole.

4. Are there locations on the Earth where you would always see the Sun throughout an entire day? If so, where?

    Near the North Pole.

Contains key information that could be included in responses. For a copy of the student handout with more space for answers, please visit www.nsta.org/solarscience.

Where Is It Night When We Have Noon?
Question Sheet

1. When it is local noon at your location, what are two other places that are also having local noon? How do you know it is noon there?

2. When it is local noon at your location, what are two places that are experiencing midnight?

3. Are there locations on the Earth where you would not see the Sun throughout an entire day? If so, where?

4. Are there locations on the Earth where you would always see the Sun throughout an entire day? If so, where?

EXPERIENCE 1.10

What Do We Think We Know? Revisited

Overall Concept

Students revisit their responses to *engage* Experience 1.1, "What Do We Think We Know?" allowing them to see how their understanding of the key concepts in the chapter has changed.

Objectives

Students will be able to demonstrate how their thinking has changed regarding

1. how the Sun's position in the sky changes throughout the day and what causes that change in position;

2. how different the motion of the Sun is in the sky for people at other locations, both elsewhere in the United States and in places outside of North America; and

3. what humans have done to deal with the differences that exist between various locations.

Procedure

1. Have students label the top of a page in their astronomy lab notebooks with "What I Know Now." Below that, have them write the first question they considered at the beginning of the unit: "How does the Sun's position in the sky change throughout the day, and what causes that change in position?" On the following page, have them write the next question at the top: "How different is the motion of the Sun in the sky for people at other locations, both elsewhere in the United States and in places outside of North America?" Finally, at the top of the third

MATERIALS

For the class:

- Whiteboard, blackboard, or poster paper where student preconceptions were recorded at the beginning of the unit

One per student:

- Astronomy lab notebook

page, have them write the last question: "What have humans done to deal with the time differences that exist between various locations?"

2. If you are not using this information to assess individual student understanding, then use the think-pair-share process described in the Introduction to this book (p. xix) to (1) have each student write how his or her thinking has changed, (2) have students discuss their comments in small groups and add more detail to their notebooks as desired, and (3) have a whole-class discussion to update the answers collected at the beginning of the unit.

3. If you want to use this information to assess each student, then you will want to collect the astronomy lab notebooks for review before the discussion in small groups. After you review the notebooks, it is still important to have the small-group and whole-class discussions that summarize what everyone learned.

Video Connections

- *Day and Night Video.* Tanya Hill, astronomer at the Melbourne Planetarium, narrates a very brief animation explaining why we have day and night on Earth (2 min.): *http://museumvictoria.com.au/ education/learning-lab/little-science/ day-and-night*

- *Young Explorers: A Brief History of Time Telling.* A production of the British Museum that includes sundials, clocks, and more (6 min.): *https://curiosity.com/rdr/ topics/history-of-telling-time*

- *Real World: Longitude and Time Zones.* A NASA eClip featuring math problems (5 min.): *https://www.youtube.com/ watch?v=kDWHM00sZJc*

- *Geography Lesson Idea: Time Zones.* A video explaining differences in time around the planet (3 min.): *www.youtube.com/ watch?v=-j-SWKtWEcU*

- *Magnetic Declination Curated by Patrick Alken.* National Oceanic and Atmospheric Administration scientist Patrick Alken discusses how rotational north differs from magnetic north and how we measure the difference, which is called magnetic declination (4 min.): *www.youtube.com/ watch?v=WwlKx96q8lE*

- *Strangest Time Zones of the World.* A history of time zones and examples of places that keep their own time (9 min.): *www.youtube.com/ watch?v=uW6QqcmCfm8*

Math Connections

- The ancient Greeks divided a circle and a sphere into 360° of angle. We can imagine that the equator of planet Earth has a circumference of 360°. When geographers and others decided to set up time zones around the Earth, they wanted to have time change by one hour per time zone. The Earth turns in one day and a day has 24 hours. How many degrees will there be in each of the 24 time zones around the Earth?

  - *Answer:* 360 divided by 24 = 15° per time zone

- At an international conference in 1884, the local time at the Royal Observatory, Greenwich, outside of London was selected as a "standard time" to begin counting the hours in the various time zones. Greenwich Mean Time eventually evolved into Universal Time (UT), which we use today. The times in the various U.S. time zones differ from UT according to the following formulas:

  - Pacific Time = UT – 8 hours

  - Mountain Time = UT – 7 hours

  - Central time = UT – 6 hours

  - Eastern Time = UT – 5 hours

(Note: When it's daylight saving time, you sub-tract one hour less than the figures given above.)

- When it is 11:00 p.m. in London, what time is it in San Francisco? What time is it in New York City? What time is it in Columbia, Missouri?

- A nice contest format for converting between time zones is found at *www.quia.com/rr/94055.html*. (See Cross-Curricular Connection 2 for more on how the time zones in our real world differ from this mathematical ideal.)

- On Mars, the day (the time the planet takes to spin once) is 24 Earth hours and 40 Earth minutes long. A Martian day is often called a sol to help distinguish it from an Earth day. Ask students (in small groups) to think about how they might devise a system of time into which to subdivide a sol for Mars. Once students have devised their own way, they can read the first few parts of the article on "Timekeeping on Mars" at *http://en.wikipedia.org/wiki/Timekeeping_on_Mars*. A computer program that shows solar time on Mars can be found at *www.giss.nasa.gov/tools/mars24*.

- NASA has a good set of math problems about Mars time keeping in their Space Math collection (go to activity 47): *www.nasa.gov/sites/default/files/files/Space_Math_IX.pdf*.

- Students who are interested in numbers and math may like to investigate the duodecimal (sometimes called dozenal) system of counting that we mention in the Content Background section of this chapter (p. 8). There is actually a Dozenal Society of America, which advocates for a system of counting based on 12 and has a website full of educational materials: *www.dozenal.org/drupal*.

Literacy Connections

Students may want to read further about the topics introduced in this chapter. Here a few places to start:

- Barnett, J. E. 1999. *Time's pendulum: From sundials to atomic clocks, the fascinating history of timekeeping and how our discoveries changed the world.* New York: Harcourt Brace.

- BBC News World. 2011. A brief history of time zones. *www.bbc.co.uk/news/world-12849630*.

- Moss, T. How do sundials work? British Sundial Society. *http://sundialsoc.org.uk/discussions/how-do-sundials-work*.

- Royal Museums Greenwich. The magnetic compass. *www.rmg.co.uk/explore/sea-and-ships/facts/ships-and-seafarers/the-magnetic-compass*.

- Social Studies for Kids. Time zones in the United States. *www.socialstudiesforkids.com/articles/time/timezones.htm*.

Cross-Curricular Connections

HISTORY OF TIME MEASUREMENT AND TIME UNITS

A fascinating topic for historical research is to learn what units of time ancient people used and how the units we use today originated. The sources below plus some of the books and websites in the Teacher Resources can get students started. As always, when students use the web for research, it's best to have them stick with websites that end in .edu, .gov, or .org.

- Atkins, W. A. 2002. The measurement of time. Encyclopedia.com. *www.encyclopedia.com/topic/Measurement_of_Time.aspx*.

- Jesperson, J., and J. Fitz-Randolph. 1999. *From sundials to atomic clocks: Understanding time and frequency.* Washington, DC: National Institute of Standards and Technology. *http://tf.nist.gov/general/pdf/1796.pdf*.

- National Institute of Standards and Technology. 1995. A walk through time: The evolution of time measurement through the ages. *www.nist.gov/pml/general/time*.

- Rogers, L. 2008. A brief history of time measurement. NRICH. *http://nrich.maths.org/6070*.

TIME ZONES AND THE REAL WORLD

Ideally, all time zones would be 15° wide, and the dividing lines between them would be great circles of longitude going from the North Pole to the South Pole and back. Everyone in a given time zone would be on exactly the same hour of standard time. However, different cities and countries have decreed or begged for exceptions to the general system. For example, the Marquesas Islands in the South Pacific have decided that time will be offset from the next zone not by one hour but by only half an hour. And Spain, although it is in the same time zone as Greenwich, England, and should be on UT, chooses instead to be on the next time zone to the east (Central European Time).

Students can each do research on one country or island that has chosen to change from the idealized standard time and then report to the class on what they have learned. For resources, they could start with the following:

- World Time Zone. Interesting and confusing facts about time / time zones. WorldTimeZone.com. *www.worldtimezone.com/faq.html*.

- Rosenberg, M. Offset time zones. About Education. *http://geography.about.com/od/culturalgeography/a/offsettimezones.htm*.

- (*For more advanced students*) Wikipedia. Time zone. *http://en.wikipedia.org/wiki/Time_zone*.

THE INTERNATIONAL DATE LINE

Time and longitude on Earth are calculated from the *prime meridian* going through Greenwich, England. This is a line connecting the North Pole, Greenwich, and the South Pole. On the exact opposite side of the world (180° of longitude away) is the *antimeridian*, where countries have agreed that the next day begins. We call the line where the date (the day of the year) changes the *international date line* (IDL; it sounds like a website for making social engagements, but that's its real name). In the ideal world, the IDL would coincide exactly with the antimeridian (they would be the same circle). But in the real world, locations near the IDL have determined that there will be exceptions, and so the line actually juts in and out to include or avoid certain territories. Students can research these exceptions and report to the class with a map of the world handy to show what's where. For resources, they could start with the following:

- U.S. Naval Observatory. 2013. The international date line. *http://aa.usno.navy.mil/faq/docs/international_date.php*.

- van Gent, R. 2008. A history of the international date line. Universiteit Utrecht. *www.staff.science.uu.nl/ ~gent0113/idl/idl.htm*.

- Wikipedia. International date line: Historical alterations. *http://en.wikipedia. org/wiki/International_Date_ Line#Historical_alterations*.

LIFE AT HIGH LATITUDES

As you go to higher latitudes on Earth, some of the phenomena discussed in this chapter become more and more dramatic. In the Northern Hemisphere during the summer, the days get longer and longer as you get to higher latitudes, and the nights get shorter and shorter. For visitors to Alaska or Scandinavia, this takes some getting used to.

Students should try to find examples in stories, art, and songs of the changes in the length of the day and the height of the Sun. For example, in the Stephen Sondheim musical *A Little Night Music*, set in Sweden, the characters complain through song that "The Sun Won't Set" even as it gets later and later in the summer evening. See the lyrics at *www.themusicallyrics.com/l/286-a-little-night- music-lyrics/2405-the-Sun-wont-set.html*.

Alternatively, you can ask students to write their own stories about what it would be like if they went to a place where the Sun only sets for a few hours out of the day or a place where the Sun doesn't set.

INVESTIGATING SUNDIALS AS CLOCKS, ART, OR ARCHITECTURE

While sundials began as the first clocks, today there are still many people making sundials privately or as public monuments. Entire organizations are devoted to sundials, and some cities point to their large sundials as a source of civic pride and tourist interest. Students can investigate some of the many uses and displays of sundials today. Here are some resources to start with:

- The British Sundial Society (also has good background info, along with much detailed and technical information): *http:// sundialsoc.org.uk*

- Frans Maes' sundial site (with pictures of many sundials around the world): *www. fransmaes.nl/zonnewijzers/welcome-e.htm*

- North American Sundial Society (a site full of materials, some introductory, many technical): *http://sundials.org*

- The Sundial Bridge in Redding, California: *www.turtlebay.org/sundialbridge*

THE CHALLENGE OF FINDING NORTH

Depending on the age of your students, you may or may not want to cover the issue of celestial north (or geographic north or rotational north) being different from magnetic north and how that difference changes on different parts of our planet (the difference is what's called magnetic declination, also known as *magnetic variation*.) This difference means that if they have a compass that points north, in most cases it won't be pointing to the north rotational pole of the Earth but only to the north magnetic pole, which is offset from rotational north and wanders as time goes on. Here are a couple of resources:

- An interactive map showing magnetic declination is available at *www. windows2universe.org/physical_science/ magnetism/north_mag_pole_interactive. html*.

- Students who want to know more about (or make a report on) how to find celestial north can read about it at *http://adventure.howstuffworks.com/survival/wilderness/true-north.htm*.

Resources for Teachers

BOOKS

- An introduction to the history of time measurement on Earth:

    Barnett, J. E. 1999. *Time's pendulum: From sundials to atomic clocks, the fascinating history of timekeeping and how our discoveries changed the world.* New York: Harcourt Brace.

- A story of how time on Earth was standardized through the use of time zones:

    Blaise, C. 2002. *Time lord: Sir Sandford Fleming and the creation of standard time.* New York: Vintage Books.

- A more historical approach to measuring time.

    Withrow, G. J. 1989. *Time in history: Views of time from prehistory to the present.* Oxford, U.K.: Oxford University Press.

- A history of time measurements and units.

    Jesperson, J., and J. Fitz-Randolph. 1999. *From sundials to atomic clocks: Understanding time and frequency.* Washington, DC: National Institute of Standards and Technology. *http://tf.nist.gov/general/pdf/1796.pdf*

- This enhanced e-book offers information about the topics in this chapter:

    National Science Teachers Association (NSTA). 2015. *Earth, Sun, and Moon.* Arlington, VA: NSTA. *https://itunes.apple.com/us/book/earth-Sun-and-Moon/id694637830?mt=11*

WEBSITES

- A website from the National Earth Science Teachers Association that is full of tools and information for teachers on Earth and space sciences:

    Windows to the Universe: *www.windows2universe.org*

- A comprehensive resource about how we keep time on Earth; has time zone converters and many other historical and mathematical tools:

    Time and Date Website: *www.timeanddate.com*

- A handy site that lets you calculate the local time and the time difference from your location in cities, airports, sporting events, and so on:

    Time Zone Guide: *www.timezoneguide.com*

- An interactive simulation that shows the motion of the Sun in the sky on any day at any latitude, letting you adjust many characteristics of what you see:

    Nebraska Astronomy Applet Project: Motions of the Sun Simulator: *http://astro.unl.edu/naap/motion3/animations/sunmotions.html*

- This site from the U.S. Naval Observatory allows you to put in your city and the date, and the calculator will show you rising and

setting times of the Sun and Moon and much more:

Sun and Moon Data for One Day: *http://aa.usno.navy.mil/data/docs/ RS_OneDay.php.*

- A history of time measurement from the National Institute of Standards and Technology:

A Walk Through Time: *www.nist.gov/ pml/general/time.*

- An article on the mathematical aspects of setting up a sundial from the American Mathematical Society:

Austin, D. 2011. The shadow knows: How to measure time with a sundial. American Mathematical Society. *www. ams.org/samplings/feature-column/ fc-2011-08.*

- The history of sundials:

Encyclopaedia Britannica. Sundial: Timekeeping device. *www.britannica.com/ EBchecked/topic/573826/sundial.*

2

Understanding and Tracking the Annual Motion of the Sun and the Seasons

I n Chapter 1, students researched the Sun's apparent daily motion and its relationship to the cardinal directions and shadows on the Earth (Figure 2.1). This chapter helps expand their thinking to how the apparent motion of the Sun in the sky changes throughout the year and the effects of those changes on the Earth, including the varying length of daytime and differences in the Sun's height in the sky. These changes on the Earth are critical for understanding what causes the seasons, which is the major thrust of this chapter.

FIGURE 2.1

The analemma, or plot, of the Sun's motion in the sky during the year, which is caused by the revolution of the Earth around the Sun. This image is produced by taking a photo of the Sun at noon every couple of weeks from the same position.

2

Learning Goals of the Chapter

After doing these activities, students will understand the following:

1. The height of the Sun above the horizon varies throughout the year in a regular and predictable pattern and is higher in the sky at noon in the summer and lower in the sky at noon in the winter.

2. The rising and setting point of the Sun changes throughout the year in a regular and predictable pattern. In the Northern Hemisphere, the Sun rises in the northeast and sets in the northwest in the spring and summer but rises in the southeast and sets in the southwest in the fall and winter. The Sun rises furthest toward the northeast and sets furthest toward the northwest on the first day of summer, and it rises furthest toward the southeast and sets furthest toward the southwest on the first day of winter. On the first day of spring and the first day of fall (the equinoxes), the Sun rises directly in the east and set directly in the west.

3. The seasons are caused by the combined effects of the tilt of the Earth's rotation axis relative to the direction of the Sun and the Earth's revolution around the Sun.

4. When the Northern Hemisphere experiences winter, the Southern Hemisphere experiences summer, and vice versa.

5. The amount of daylight any place on Earth experiences in a given season depends on its latitude.

6. Our common measures of time, the day and year, are determined by the rotation rate of the Earth and the revolution period of the Earth around the Sun.

Overview of Student Experiences

ENGAGE EXPERIENCES

This chapter begins with what many students see as a contradiction to their current thinking: How can it be summer on one part of the Earth when it is winter elsewhere? The goal is to hook them into wanting to learn more about how the changing apparent motion of the Sun in the sky can resolve this contradiction.

- **2.1. What Do We Think We Know?** Students discuss and record their thoughts regarding the following questions: (1) Does the Sun's position in the sky at noon change throughout the year? (2) If so, how does it change, and what causes the change? And (3) what causes the differences we experience with the changing seasons, such as length of daylight and temperature?

- **2.2. How Can This Be True?** Students read an e-mail and look at a photo sent from a friend or relative in the Southern Hemisphere. The photo shows people celebrating the New Year on the beach—clearly enjoying the sunny, hot weather. Students also have a second e-mail and photo taken at the same day and time from a friend or relative in the Northern Hemisphere who complains of the freezing cold. Students discuss and record their thoughts regarding why it is so warm in one place while it is so cold in another on the same day of the year.

EXPLORE EXPERIENCES

Now that students want to understand what causes the differences observed in the Northern Hemisphere and Southern Hemisphere, the following two activities build on the experiences in Chapter 1 to have them collect and analyze data over a number of months. This leads them to the data and understanding needed to deal with the physical and mental models in the Explain section of the chapter.

- **2.3. Sun Tracking Throughout the Year:** This activity continues Experience 1.3, "Shadow and Sun Tracking," from Chapter 1 so that students have data from multiple months. Students use the data to determine how much the height and direction of the Sun changes over the course of the year.

- **2.4. High Noon Throughout the Year:** Students use the "Shadow and Sun Tracking" data to determine the height of the Sun at local noon in different months of the year. They conclude that the Sun is highest in the sky at noon during the summer and lowest in the sky at noon in the winter.

EXPLAIN EXPERIENCE

Students use the data collected in the prior experiences to develop and use physical and mental models to understand how the Sun's daily motion across the sky changes throughout the year and to grasp the fundamental causes of the seasons we experience on Earth.

- **2.5. Reasons for the Seasons Symposium:** Students work in small groups to explore factors that might be causes of the seasons: Earth's changing distance to the Sun, the amount of daylight each day at different times of year, and the angle of the Sun above the horizon. After they do their independent research on the factors, they come together in a "symposium" to present their results and reach appropriate conclusions.

ELABORATE EXPERIENCES

Students engage in additional activities to reinforce and practice using the physical and mental models developed in previous experiences.

- **2.6. Length of Day Around the World:** This activity builds from Experience 1.5, "Noontime Around the World." Gnomons are attached to the globe, with one in the Southern Hemisphere, one in the Northern Hemisphere, and one on the equator. Students observe how the amount of daylight and the gnomon's shadow vary depending on the gnomon's latitude and the season.

- **2.7. Seasons on Other Planets:** Students learn that on other planets both the varying distance from the Sun (eccentricity of orbit) *and* the tilt of the planet's axis can influence the seasons. Then they look at characteristics of a few other planets and try to predict what the seasons on them are like.

- **2.8. I Can't Make It Come Out Even: Fitting Days and Years Into a Workable Calendar:** Students are told the length of an Earth year and an Earth day and are challenged to make a consistent calendar. After learning about the Julian

and Gregorian calendar reforms for our planet, they are given data for an imaginary planet, Ptschunk, and have to apply their knowledge to create a calendar for that world.

EVALUATE EXPERIENCES

These experiences allow the teacher to assess how well the students understand the key ideas presented and can demonstrate the scientific practices developed by doing the experiences in the chapter. We would not expect teachers to do all of the *evaluate* experiences but to choose the ones that best fit their learning goals for the students and the integration strategies they use (e.g., the first experience connects well with language arts skills desired in students).

- **2.9. Write a Picture Book for Kids:** Each student writes and illustrates a picture book for a younger student that explains how the relationship between the Earth and Sun causes the seasons.

- **2.10. E-mail Response to "How Can This Be True?":** Each student prepares an e-mail response to the friend in the *engage* experience at the beginning of this chapter that explains why the Southern Hemisphere can experience summer while the Northern Hemisphere experiences winter.

- **2.11. Reasons for the Seasons Revisited:** Students label a diagram that shows the relationship of the Sun and Earth at different times of year and describe how this explains the reasons for the seasons.

- **2.12. What Do We Think We Know? Revisited:** Students revisit their responses to the questions from the first *engage* experience to discuss how their thinking has changed after experiencing the activities in the chapter.

Recommended Teaching Time for Each Experience

Table 2.1 provides information about the teaching time needed for each experience in this chapter. Please note that some of the experiences require sunny days. These are indicated in the table.

Connecting With Standards

Table 2.2 (pp. 72–73) shows the standards covered by the experiences in this chapter. This book does not deal with aspects of the *Next Generation Science Standards* performance expectations that involve the stars at night. Chapter 1 deals with changes in length and direction of shadows made by the Sun during a single day and its relationship to the rotation of the Earth and the observer's location on the Earth. Chapter 4 addresses lunar phases and eclipses. Additionally, this table does not give *Common Core State Standards* but gives general connections to the language arts and mathematics standards.

TABLE 2.1

Recommended teaching time for each experience

Experience	Time
Engage experiences	
2.1. What Do We Think We Know?	**45 minutes**
2.2. How Can This Be True?	**30 to 45 minutes**
Explore experiences	
2.3. Sun Tracking Throughout the Year ●	Assuming the students did "Shadow and Sun Tracking" in Chapter 1, then **10 to 15 minutes** are needed for each **five or six observations** made throughout the day; **30 minutes** for follow up discussion, typically on the next day
2.4. High Noon Throughout the Year ●	Should be done as part of "Shadow Tracking Throughout the Year"; adds **30 minutes** to observations at local noon
Explain experiences	
2.5. Reasons for the Seasons Symposium	**Four 45-minute periods**
Elaborate experiences	
2.6. Length of Day Around the World	**45 to 60 minutes**
2.7. Seasons on Other Planets	**45 to 90 minutes**
2.8. I Can't Make It Come Out Even: Fitting Days and Years Into a Workable Calendar	**Two periods of 45 minutes** each; additional time for the optional extras
Evaluate experiences	
2.9. Write a Picture Book for Kids	**120 minutes** over several days, or do a portion as homework
2.10. E-mail Response to "How Can This Be True?"	**30 to 45 minutes**
2.11. Reasons for the Seasons Revisited	**30 minutes**
2.12. What Do We Think We Know? Revisited	**30 to 45 minutes**

● = Requires a sunny day

TABLE 2.2

Chapter 2 *Next Generation Science Standards* and *Common Core State Standards* connections

Performance expectations	• 5-ESS1-2: Represent data in graphical displays to reveal patterns of daily changes in the length and direction of shadows, day and night, and the seasonal appearance of some stars in the night sky. • MS–ESS1-1: Develop and use a model of the Earth-Sun-Moon system to describe the cyclic patterns of lunar phases, eclipses of the Sun and Moon, and seasons.
Disciplinary core ideas	• 5-ESS1.B: (Earth and the Solar System): The orbits of Earth around the Sun and of the Moon around Earth, together with the rotation of Earth about an axis between its North and South Poles, cause observable patterns. These include day and night; daily changes in the length and direction of shadows; and different positions of the sun, moon, and stars at different times of the day, month, and year. • M-ESS1.A: (The Universe and Its Stars): Patterns of the apparent motion of the Sun, the Moon and stars in the sky can be observed, described, predicted, and explained with models.
Science and engineering practices	• Develop and use a model to describe phenomena (e.g., model of the tilted Earth orbiting around the Sun and the implications for the seasons experienced at different locations on the Earth). • Analyze and interpret data to determine similarities and differences in findings (e.g., analysis and interpretation the amount of sunlight reaching the Earth depending on the angle of the Sun above the horizon). • Engage in argument from evidence (e.g., discussion of what factors are the primary cause of the Earth's seasons in the "Reasons for the Seasons Symposium" experience).

Content Background

THE SUN'S MOTION OVER THE COURSE OF A YEAR AND THE SEASONS

Astronomers believe that early in its history the Earth had one or more accidental encounters with large objects that were—in those early days—part of the solar system. As a result, the Earth no longer orbits the Sun with its *axis* (the line connecting the north rotational pole with the south rotational pole) vertical. Instead, our axis became tilted by about 23.5°, and like many an accident victim, the Earth has no way of straightening up. This tilt of our axis has many consequences for

Table 2.2 (*continued*)

Crosscutting concepts	• Patterns can be used to identify cause-and-effect relationships (e.g., analysis of the variation in the shadow of the sun clock gnomon throughout the year and its relationship to Sun's path across the sky).
	• Science assumes that objects and events in natural systems occur in consistent patterns that are understandable through measurement and observation (e.g., observations related to the changing path of the Sun across the sky throughout the year reveals patterns associated with the Earth's seasons).
	• Models can be used to represent systems and their interactions (e.g., use of the Earth–Sun model to understand the primary causes of the Earth's seasons).
Connections to the *Common Core State Standards*	• *Writing:* Students write arguments that support claims with logical reasoning and relevant evidence, using accurate, credible sources and demonstrating an understanding of the topic or text. The reasons and evidence are logically organized, including the use of visual displays as appropriate.
	• *Speaking and listening:* Students engage effectively in a range of collaborative discussions (one-on-one, in groups, and teacher-led) with diverse partners, building on others' ideas and expressing their own clearly. Report on a topic or text or present an opinion, sequencing ideas logically and using appropriate facts and relevant, descriptive details (including visual displays as appropriate) to support main ideas or themes.
	• *Reading:* Students quote accurately from a text when explaining what the text says and when drawing inferences from the text. Students determine the meaning of general academic and domain-specific words and phrases in a text relevant to the student's grade level.
	• *Mathematics:* Students recognize and use proportional reasoning to solve real-world and mathematical problems. Students summarize numerical data sets in relation to their context, including reporting the number of observations and describing the nature of the attribute under investigation, including how it was measured and its units of measurement.

life on Earth as we orbit the Sun, and we explore them in this chapter.

It's important to remember that this tilt is a physical leaning of our planet in space in one particular direction. The North Pole of Earth continues to point in that same direction during the entire year, as we move from one side of the Sun to the other. In June, the North Pole of the Earth leans in toward the Sun, while the South Pole leans away (Figure 2.2, p. 74).

FIGURE 2.2

The orientation of the Earth during the four seasons

The North Pole points in the same direction for the whole year.

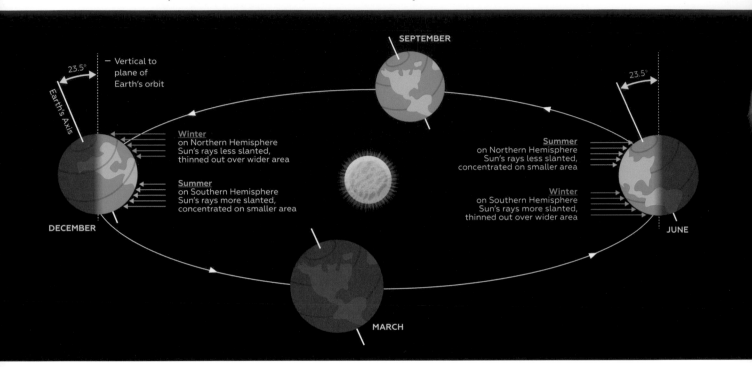

This means that in June sunlight hits the Northern Hemisphere more vertically, while it hits more obliquely in the south. You can get this same effect by shining a flashlight on a wall at night. When the flashlight is pointed straight at the wall (when the rays point more vertically), the spot of light is more intense, but if you shine the beam at an oblique angle, the spot spreads out, and there is less light at any given place on the wall.

Another effect of the Earth's tilt is that, in the Northern Hemisphere, the Sun moves higher in the sky in June during the course of a day. The Sun also rises north of east in June and sets north of west, spending more time with those of us in the Northern Hemisphere compared with those in the Southern Hemisphere. Thus, the northern

days are longer (while the nights are shorter). Notice how these two effects reinforce each other in June: Just as the Sun is more effective at heating (the rays are more vertical) in this month, the Northern Hemisphere provides it with more time to heat and less time to cool off (longer days and shorter nights).

The effect is reversed in the south. The Sun's rays in June are less effective at heating. At the same time, the Sun never goes very high in the sky, and the days are shorter and the nights longer. This means that, just as the Sun is not very good at heating, it also has less time to heat up the Southern Hemisphere and more time (at night) to cool it off. The south therefore has winter in June.

74

In March and September, the Earth's tilt is sideways to the Sun, so both hemispheres get equal treatment. The angle at which sunlight comes in is the same, and the days and nights are of equal length in each hemisphere.

In December, the situation is reversed from June. At this time, it is the Northern Hemisphere that leans away from the Sun and the Southern Hemisphere that leans toward the Sun. Thus, the rays of the Sun are more effective at heating in the south just when the days are long and the nights are short. So the Southern Hemisphere has summer in December.

In the Northern Hemisphere, on the first day of winter (roughly December 22), the Sun at noon will reach its lowest altitude in the sky for the year. The sun rises south of east and sets south of west, so it spends less time in the sky with northerners. The days are shorter, and the Sun's rays are more spread out. Anyone who wants to have long, sunny days in December has to fly toward our planet's southern half.

THE SUN'S MOTION AND THE SEASONS AT DIFFERENT LATITUDES

The seasonal effects depend very much on your latitude on Earth (latitude being how far north or south you are from the equator of our planet). At the equator itself, seasonal effects are pretty minimal. The Sun is up for 12 hours and down for 12 hours, and the presence or absence of rain has a greater effect on the seasons than sunlight.

The more you move away from the equator, the more pronounced the seasonal effects become. At the latitude of U.S. cities such as Boston or Chicago, the differences between summer and winter are quite noticeable. In the summer, the Sun is high in the sky in the middle of the day, the days are long, the Sun's light is intense, and the heat leads to the use of air conditioners in many homes and offices. In the winter, the Sun stays low in the sky in the middle of the day, the days are short, and the Sun's light is more diffuse and less effective at heating. Thus, people have to turn on their heaters to keep themselves warm.

When you get really far north, the seasonal differences become downright dramatic. Visitors to Alaska and northern Canada marvel at the extremely long days and short nights in summer. Places within 23° of latitude of the North Pole have a period of time when the Sun never sets. At the North Pole, the Sun is simply up all the time from late March to late September. At the South Pole the Sun doesn't set from late September to late March.

GETTING THE REASONS FOR THE SEASONS RIGHT

We should note that the reasons for the seasons is one of the most misunderstood topics in science education. The following website lists some of the misconceptions about the seasons that are commonly seen among students: *www.lpi.usra. edu/education/workshops/phasesSeasons/ SeasonsMisconceptions.pdf.*

Surveys of adult Americans show that many people think that it is the Earth's changing distance from the Sun that causes the seasons. It's not surprising that students also have difficulty with this question. By making sure that our students really internalize their understanding of the seasons in school, we can help see to it that the coming generations are better informed on this issue than their parents.

On the issue of the effect of changing distance from the Sun on the seasons, it turns out the

Earth is closest to the Sun in January and farthest from the Sun in July. The difference between those maximum and minimum values and the average distance is less than 3%, so this is not a significant effect. Students in the United States who think distance is the crucial factor for the seasons are often shocked when you ask them to think about the seasons in Australia. If distance governs the seasons, how could Australia have the opposite seasons from us? After all, we are all on the same planet and the same distance from the Sun at any given time.

As students will discover when they participate in the experiences in this chapter, the main factors in the seasons for our planet are the consequences of the tilt of our axis.

SEASONS ON OTHER PLANETS

Other planets may or may not have tilted axes like the Earth. Of our two neighbor planets, for example, Venus has no tilt and no seasons, while Mars has a tilt roughly the same as Earth's and shows pronounced seasonal effects.

In addition, some planets have a more significant difference in their distance from the Sun over the course of their year than the Earth does. The *eccentricity* of an orbit is a measure of how close to a perfect circle (equal distances from the center) the path of a planet is (Figure 2.3). Eccentricity varies from 0 (meaning a circle) to 1 (meaning the orbit would be stretched infinitely and become a line). The more elongated (elliptical) an orbit is, the greater the eccentricity.

You can see the axis tilt and eccentricity of each planet in our solar system in the "Characteristics of the Planets" handout (p. 129) in Experience 2.7,

FIGURE 2.3

An illustration of one planet with small eccentricity and one with large eccentricity

"Seasons on Other Planets." Mars is an example of a planet for which both the tilt and the eccentricity play a role in the seasons and, as a result, the seasons have different intensity in the Northern and Southern Hemispheres.

As of the time we are writing this book, astronomers have discovered over 2,000 planets orbiting other stars. While we do not yet know enough about those planets to understand their seasons, it's fair to say that we expect to discover even more extreme and interesting seasonal differences among them.

HOW CELESTIAL MOTIONS GIVE US OUR MEASURES OF TIME

Two of our basic units of time measurements are based on the movement of the Earth through space. The Earth's rotation on its axis is the length of our day, and its revolution around the Sun gives us the length of our year. (The month is derived from the cycle of the Moon, which we will discuss in Chapter 4.)

There are roughly 365.25 days in a year, which is one of the challenges in producing a calendar that continues to work for centuries. Many cultures

TABLE 2.3

Key days for marking the seasons

Date	Season	Name	Characteristics in the Northern Hemisphere
June 21	Summer begins	Summer solstice	Longest day, shortest night
September 22	Fall begins	Fall equinox	Day and night are equal
December 22	Winter begins	Winter solstice	Shortest day, longest night
March 21	Spring begins	Spring equinox	Day and night are equal

based their calendars on the cycle of the Moon's phases, which is a lot easier to keep track of (see next section, Lunar Calendars). The Romans, under Julius Caesar, introduced the notion of a leap year, so that every four years we add an extra day to take care of that pesky quarter of a day.

However, the year is not exactly 365.25 days long, and the so-called Julian calendar (named for Julius Ceasar), was off by 11 minutes per year from the true value. As the centuries went on, the 11 minutes started to add up, and the seasons and the months began to get out of alignment. (By the 1500s, spring was starting March 11 instead of March 21!) So, under Pope Gregory in the 1580s, the Gregorian calendar made further adjustments, with only century years divisible by 400 being leap years.

The key days for marking the seasons are shown in Table 2.3. The term *equinox* means "equal night." It is the time when the day and the night are the same length. The word *solstice* means "Sun standstill." On those dates, the Sun reaches its extreme rising and setting positions—farthest to the north or farthest south. The Sun is also highest in the sky or lowest in the sky at noon on those dates. It seems at that point to momentarily stand still and then reverse course.

LUNAR CALENDARS

As noted earlier, the most practical calendar for ancient people was one that relied on the regularly repeating cycle of the phases of the Moon, which repeats every 29.5 days. Since the Moon is easily visible and its cycle can be seen by everyone, this is a calendar that is both easier to share and easier to remember than one involving a full year of tracking the altitude of the Sun or the length of daylight. In a lunar-cycle based calendar, each month begins when the smallest crescent Moon that your eyes can see appears in the sky. Since 29.5 days doesn't divide evenly into 365.25, the months of a lunar calendar get out of step with the cycle of the seasons that makes the year.

To be precise, 12 *lunations* (repetitions of the lunar phase cycle) equal 354.37 days, while the solar year (sometimes called the tropical year) is 365.24 days. Various calendar systems use different rules to bring things back into step periodically. For example, in the Hebrew calendar, an extra month is added seven times every 19 years. This is why the holidays in some cultures that are calculated using the lunar calendar happen at different times in different years of our solar calendar. Such holidays include Chinese New Year, Ramadan, and Hanukkah.

EXPERIENCE 2.1

What Do We Think We Know?

Overall Concept

Just as with the first activity in Chapter 1, this beginning experience allows students to reflect on what they already know about the subject in the chapter. This gives you an understanding of their thinking and identifies any preconceptions that need to be dealt with.

Objectives

Students will

1.   reflect on and write in their astronomy lab notebooks what they think they know about how the daily motion of Sun in the sky changes over the year and how that relates to the seasons and

2.   share their ideas with other students.

Advance Preparation

Identify space where the list of student preconceptions can be located for the length of time the class is researching the Sun's annual motion. Copies of the photos and e-mail templates are available at *www.nsta.org/solarscience*.

Procedure

1.   Tell students that now that they have explored how the Sun's position in the sky varies in a single day, you would like to understand what they already know about how the Sun's position and motion in the sky changes throughout the year.

MATERIALS

For the class:

*   Whiteboard, blackboard, or poster paper where student preconceptions can be recorded and remain visible for the duration of the time spent researching the Sun's motion throughout the year

One per student:

*   Astronomy lab notebook

2. Have them label the top of a page in their astronomy lab notebooks with "What I Think I Know." Right below that, have them write the first and second question you want them to consider and discuss: "Does the Sun's position in the sky at noon change throughout the year? If so, how does it change, and what causes the change?" At the top of the following page, have them write the next question: "What causes the differences we experience with the changing seasons, such as length of daylight and temperature?"

    Teacher note: If your school is in a location where the differences between seasons are not extreme, you may want students to think about what they know about seasons in places such as Boston and Chicago.

3. Use the think-pair-share process described in the Introduction of this book (p. xix) to (1) have students individually write answers to the questions in their astronomy lab notebooks; (2) discuss their answers with other students in small groups and add more detail to their notebooks as desired; and finally, (3) come together for discussion with the whole class, report their conclusions, and produce a list of what they think they know in a location in the classroom where the information can be kept for future reference.

4. Finish up the experience by thanking the students for sharing what they think they know and emphasizing that keeping the notes up in the classroom will help the class see what new things they learn during the study of the Sun's motion and position throughout the year.

EXPERIENCE 2.2

How Can This Be True?

Overall Concept

Students read what appear to be contradictory e-mails about the weather on the same day at two different places on Earth and begin thinking about what causes the seasons.

Objective

Students will realize that weather at two locations can be dramatically different, even if the local environments are similar, and start to consider reasons for this difference.

Advance Preparation

Produce (and personalize if you wish) the two e-mails using the templates provided below, so that they appear to be e-mails from a friend. Print copies for each student. Copies of the photos and e-mail templates are available at *www.nsta.org/solarscience*.

Procedure

1. Distribute the two e-mails to students, telling them that you received these messages from a friend, who received them from two relatives. Your friend cannot understand how the conditions can be so different in the two locations described in the e-mails. Tell the students that you want to provide a response to your friend and would like to get suggestions from your students for what to say. (If the class doesn't believe the part about the Santa Run in Australia, you can show them this video: *www.youtube.com/watch?v=NP-H0MrpolE*.)

MATERIALS

One per student:

- Simulated e-mails from two friends in two different locations, including Figures 2.4 and 2.5 (pp. 82–83)

- Astronomy lab notebook

2.2

ENGAGE

2. Use the think-pair-share strategy (discussed in this book's Introduction, p. xix) to have students write ideas regarding what explains the differences described in the e-mails, carry out small-group discussions, and finally have a discussion as a whole class.

3. Save the ideas in the whole-group discussion so you can return to them after you complete the experiences in this chapter.

Teacher note: The goal at this time is not for students to provide the "right" answers but for you to identify the range of answers and preconceptions that your students have regarding seasons and other factors that may affect weather at a given place on the Earth.

FIGURE 2.4

Sample e-mail for the "How Can This Be True?" experience
(Feel free to modify as desired for your specific situation.)

INBOX

From Brooklyn, New York:

Dear (friend's name),

I would say, "I wish you were here," but I'm not sure why anyone would want to be here in the −10° weather with 18 inches of snow on the ground—see photo. I did venture out briefly today to see some holiday lights, but couldn't stay out very long in the cold.

I hope your holiday season is more pleasant.

Peggy (or other name of choice)

FIGURE 2.5

Sample e-mail for the "How Can This Be True?" experience
(Feel free to modify as desired for your specific situation.)

INBOX

From Sydney, Australia:

Dear (friend's name),

Having a wonderful time enjoying the holidays in Sydney. Spent the day at a barbecue on Bondi Beach with thousands of other people—see attached photo. Then went to Circular Quay where we watched the Santa Fun Run in 90° weather—see the second photo. Not sure how they can stand to be in those Santa costumes in such heat.

Hope you're having a wonderful festive season.

Deb (or other name of choice)

EXPERIENCE 2.3

Sun Tracking Throughout the Year

Overall Concept

This experience continues Experience 1.3, "Shadow and Sun Tracking." Students can use the data to determine how much the height and direction of the Sun in the sky changes throughout the year. Ideally, the data are collected every two months over at least six months, but you can do this in a shorter period of time by using data you or other classes collected at other times of year (see specific suggestions under Advance Preparation).

The shadow of the gnomon on the toilet plunger sundial follows the same general pattern throughout the year, but the length of the shadow at any given time is different from what it was in the observations made in Chapter 1. The direction of the shadow is the same around noon but can be different at times earlier or later in the day.

The height of the Sun above the horizon at noon in the Northern Hemisphere also varies throughout the year. It is at the highest on the summer solstice (around June 21) and lowest on the winter solstice (around December 21).

In the Northern Hemisphere, the Sun rises in the northeast and sets in the northwest in the spring and summer, and rises in the southeast and sets in the southwest in the fall and winter. On the first day of spring and fall (the equinoxes) the Sun rises directly in the east and sets directly in the west.

Objectives

Students will

1. observe the changing direction and length of a shadow over a portion of a day and compare the direction and length to observations in prior months,

MATERIALS

One per group of three to four students:

- A small toilet plunger (a cheap wood-handled plunger works well; the handle should be cut to about 22 cm so that the shadow will fit on your poster-sized sheet of paper)

- Chalk and a black marker

- Poster-sized sheet of blank paper

One per student:

- "Where Is the Sun Now? Observation Diagram Sheet" (p. 29)

- "Sun Tracking Throughout the Year Discussion Questions" (p. 89)

- Astronomy lab notebook

2. determine how the changes observed in the shadow relate to the motion of the Sun across the sky,

3. determine how the height of the Sun above the horizon at noon changes throughout the year, and

4. determine how the change in shadow direction and length relates to the time of year and the season.

Advance Preparation

1. Ideally, this experience is done every two months over at least six months after the students have made their observation in Experience 1.3, "Shadow and Sun Tracking."

2. The experience requires that it be sunny for as much of the day as possible, so be aware of the weather forecast as you plan for this activity.

3. The observations should include as much of the day as possible, which requires careful planning for students who have complex schedules. Have the teams do their initial setup in the morning and record the shadow length every hour until 11:00 a.m. (standard time). They then need to record the shadow every 5 to 10 minutes, as noted in the Procedure, and then each hour throughout as much of the day as possible. The more data collected throughout the day, the better your students will understand the pattern of the Sun's motion.

4. If you teach multiple classes throughout the day, you will need to be creative to complete this experience. If you have a number of classes doing the same activity, have them share their observations of the gnomon's shadow, with classes early in the day providing morning observations and those later in the day adding noontime and afternoon observations on the same poster paper. If you teach only one or two classes during the day, identify students who are willing to go out during lunch or other free time to make the observations for the entire class.

5. If you are not able to sustain doing this unit over much of the school year, then a good alternative is for you to make your own observations every few months before starting Experience 1.3,

"Shadow and Sun Tracking Shadows," from Chapter 1. You can then have the students use copies of your observations to see the change in gnomon shadow length and direction throughout the year. If you choose this option, try to get observations close to the solstices and equinoxes.

6. Please recall that if you are on daylight saving time, your clock time is off from standard time by one hour. Subtract one hour from what your clock shows (i.e., 1:00 p.m. is actually noon).

Safety note:
Emphasize to your students that they will be making an indirect observation of the Sun. They should never look directly at the Sun because even a brief viewing can damage the retinas of their eyes.

Procedure

1. Before going outside, have students review any toilet plunger sundial data they previously collected. Have them describe what they learned about the length and direction of the gnomon's shadow.

2. Give students a new piece of poster paper and have them follow the directions in steps 1–7 in Experience 1.3, "Shadow and Sun Tracking," from Chapter 1 (pp. 25–27). By the end of the day, they should have a set of observations to compare with the those made in Experience 1.3.

3. Repeat the observations after another couple of months. If you cannot sustain this study over at least six months, then share the observations you made using the same equipment over the prior six months. See alternative approaches suggested under Advance Preparation.

4. Each time the students make their gnomon shadow observation, they should also record their observations on a "Where Is the Sun Now? Observation Diagram Sheet" (see p. 29 in Chapter 1). This should be done as homework in which students find a place where they can mark the location of the Sun soon after sunrise and shortly before sunset. Again, if you cannot sustain this study over six months, then use the observations you made from the same location during the previous six months.

5. Once the students have observations from at least three different dates over a minimum of six months, they should write answers in their astronomy lab notebooks to the "Sun Tracking Throughout the Year Discussion Questions" and consider them in their teams. Then have a whole-class discussion and record the consensus to the answers on the board for later reference. As the students talk about their answers to the questions, be sure to ask them to back up their conclusion with evidence based on their observations.

6. At this time, it is not necessary to resolve any differences of opinion, but let the students argue their points of view based on the evidence they collected. Resolution of differences should occur as you work with students with the *explain* experiences in the next section.

7. (*Optional*) Consider having students plot the length of the shadow (*y*-axis) versus the time of day (*x*-axis) from their gnomon shadow observations. If they plot the length of the shadow at several times of year, they will see the pattern in how the length of the day and the height of the Sun above the horizon varies. This activity will also develop and use the mathematical thinking skills identified in the *Common Core State Standards* (i.e., students should represent real-world problems by graphing points and interpreting the resulting graph in the context of real-world situations).

Sun Tracking Throughout the Year
Discussion Questions

TEACHER VERSION*

1. Describe the differences you observed between the gnomon's shadow at a given time at different times of the year. Be sure to include a discussion of the following:

    *   The length of the shadow at several times of day

    *   The direction of the shadow at several times of day

    In the fall and winter, the gnomon's shadow was longer than at the same time in the spring and summer. The direction of the shadow at noon is always the same (pointing north), but the shadows are usually a different length and direction when you get further away in time from noon.

2. How does the movement of the Sun in the sky differ from one time of year to another? *The Sun gets higher in the sky in the spring and summer compared with fall and winter. The Sun rose in the northeast and set in the northwest in the spring and summer. The Sun rose in the southeast and set in the southwest in the fall and winter. On the equinoxes, it rose exactly in the east and set exactly in the west.*

    Teacher note: This answer assumes that students examined observation data for the entire year. If they only looked at six months of data, a partial answer is reasonable.

3. What did you observe about the length of time that the Sun was above the horizon (the amount of daytime) at different times of year? *Daytime lasts longer in the spring and summer compared with fall and winter.*

    Teacher note: Recall that the day and night are equal on the first day of spring and the first day of fall. After that, the days get longer in spring and shorter in fall.

*Contains key information that could be included in responses. For a copy of the student handout with more space for answers, please visit www.nsta.org/solarscience.

Sun Tracking Throughout the Year
Discussion Questions

1. Describe the differences you observed between the gnomon's shadow at a given time at different times of the year. Be sure to include a discussion of the following:

   • The length of the shadow at several times of day

   • The direction of the shadow at several times of day

2. How does the movement of the Sun in the sky differ from one time of year to another?

3. What did you observe about the length of time that the Sun was above the horizon (the amount of daytime) at different times of year?

EXPERIENCE 2.4

High Noon Throughout the Year

Overall Concept

This experience builds from Experience 1.7, "High Noon," in Chapter 1. Students use simple geometry and the law of similar triangles to determine the height (in degrees) of the Sun above the horizon at local noon. The height of the Sun above the horizon at noon changes throughout the year, reaching its highest point in the sky on the summer solstice and its lowest point (relative to noon during the rest of year) on the winter solstice. This demonstrates that the Sun's change in position provides a discernable pattern, which can be used to determine a person's latitude on the Earth.

Objectives

Students will

1. measure the altitude of the Sun in the sky around local noon at different times of the year and

2. analyze the data to determine the pattern for how the altitude of the Sun at noon changes throughout the year.

Advance Preparation

1. Students need to measure the altitude of the Sun around solar noon, so they will need to be outside for about 40 minutes. The time needed to take the reading can be decreased if you are able to determine exact solar noon for your location using data from the internet. A solar noon calculator is also available (*www.esrl.noaa.gov/gmd/grad/solcalc*).

MATERIALS

One per group:

*   Large sheet of blank newsprint or chart paper (at least 45 cm on one side)

*   Meterstick

*   Gnomon (toilet plunger) from prior experiences

One per student:

*   Small plastic protractor

*   "High Noon Data Chart" from Experience 1.7 (p. 50)

*   Astronomy lab notebook

2. If you are not able to sustain doing this experience with the students over much of the school year, then a good alternative is for you to make your own observations for a number of months before you start and to share those with students. You can then have the students use copies of your observations to calculate the changing altitude of the Sun at local noon. If you choose this option, try to get observations close to the solstices and equinoxes.

3. Please recall that if you are on daylight saving time, your clock time is off from standard time by one hour. Subtract one hour from what your clock shows (i.e., 1:00 p.m. is actually noon).

Safety note: Emphasize to your students that they should be making an indirect observation of the Sun. They should never look directly at the Sun because even a brief viewing can damage the retinas of their eyes.

Procedure

1. Students mark the end of the gnomon's shadow every five minutes for as long as it takes to identify which measurement shows the shortest shadow. This is when the Sun is highest in the sky and it is local noon. The length of the shadow and the length of the gnomon should be entered into the "High Noon Data Chart."

2. Students should follow the procedures given in the steps 2–4 of Experience 1.7, "High Noon" (p. 48), to calculate the altitude of the Sun to enter into the data chart in their astronomy lab notebooks.

3. Ideally, measurements of the Sun's altitude in the sky will be collected every couple of weeks or at least every month for as much of the school year as possible. It is especially important to get observations around the solstices and equinoxes. See alternatives provided in Advance Preparation if the experience cannot be sustained over an extended period of time.

4. (*Optional*) Similar to the graphing activity in Experience 2.3, students can plot the altitude (in degrees) of the Sun at noon (*y*-axis) versus the date (*x*-axis). If you were able to collect data for an entire year, this graph will show that the Sun is higher in the sky in spring and summer and lower in the sky in fall and winter and also reveal that the height of the Sun in the sky is the same on the

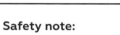

two equinoxes. As noted before, producing this graph will develop and use the mathematical thinking skills identified in the *Common Core State Standards* (i.e., students should represent real-world problems by graphing points and interpreting the resulting graph in the context of real-world situation).

5. Conclude the experience by having students write in their astronomy lab notebooks what conclusions they can make from the data collected using the following three questions (*key information that could be included in responses is in italics*):

    a. What pattern can you observe in the changing height of the Sun throughout the year? *From winter to summer, the Sun gets higher in the sky each day at noon. From summer to winter, the Sun gets lower in the sky each day at noon.*

    b. What time of year is the Sun highest in the sky at noon? What time of year is it lowest? *The Sun is highest in the sky at noon in summer, reaching its highest altitude on the summer solstice. The Sun is lowest in the sky at noon in winter, reaching its lowest altitude on the winter solstice.*

    c. How do you think this information might change for various people at other locations on the Earth (e.g., people who live in northern Canada or southern Mexico)? How can this information help a person determine the latitude of his or her location on the Earth? *As one changes latitude, the altitude of the Sun also changes. If we know the date and time at a given location, we can use the height of the Sun above the horizon to determine our latitude.*

        Teacher note: This third question is a thought question to get students thinking about how the observations may vary for other locations on the Earth. If you choose to do it, it is a good lead-in to the first *elaborate* experience. For students who have not learned about latitude and longitude, you may need to explain those concepts to them before they can answer the question.

Follow up individual writing time with small-group discussion and then finish with a whole-class conversation.

EXPERIENCE 2.5

Reasons for the Seasons Symposium

Overall Concept

Students work in small groups to explore different factors that might contribute to the differences in light and temperature during the seasons. These include Earth's changing distance to the Sun, the amount of daylight each day at different times of year, and the angle of the Sun above the horizon. After the students do their independent research, the groups come together in a "symposium" to present their results and reach conclusions: The changing distance to the Sun does not contribute to the seasonal changes, while the key factors are the angle of the Sun above the horizon and the amount of daylight. The last two factors are caused by the tilt of the Earth's axis relative to the Sun as our planet orbits the Sun over the course of a year.

Objectives

Students will

1. understand some of the processes and practices that scientists use to study a problem (e.g., identify a problem, design experiments, conduct experiments, interpret data, and present results);

2. brainstorm about various factors that could produce the Earth's seasons;

3. research the effect of the various factors that could influence the Earth's seasons;

4. conduct a symposium to present their conclusions based on their research;

5. understand that the key factors influencing the Earth's seasons are the angle of the Sun above the horizon and the amount of daylight, caused by the combination of the tilt of the Earth's axis relative to the Sun and the Earth orbiting the Sun; and

6. understand that the changing distance to the Sun does not cause the seasons.

MATERIALS

The following list includes all the materials required for the three areas of investigation in the experience. The list is based on six research groups with two groups for each area of investigation. A materials list for each area of investigation is also provided just before the instructions for that investigation.

*   4 short lamps (simple ceramic lightbulb holders work well if the electrical connections are well insulated)

*   2 250 W lightbulbs (*Safety note:* This lightbulb will get very hot. Caution students not to touch it.)

*   2 40 W lightbulbs

*   2 6 ft. (or longer) measuring tapes (preferred) or 2 metersticks

(continued on p. 94)

Advance Preparation

Surveys reveal that not just children but also most adult Americans are not familiar with the real reasons for the seasons. The most common misconception is that seasons are caused by the distance of the Earth from the Sun—that it is farther away in winter and closer in summer. Even when students are told the true reason (in textbooks or lectures), they soon return to their preconception that distance is the main cause. The best way for them to learn the reasons for the seasons is to experiment with the factors themselves and then discuss their results with their peers.

This lesson will take about four regular class periods to complete. Students first discuss possible reasons for the seasons, coming up with a list of suggested factors. They are then assigned to research groups to test the importance of some of these factors. In an all-class symposium, the groups report back to the whole class about their results. All the students have a chance to discuss and refine their ideas on the basis of experiments the groups have done. This models the way groups of scientists often approach problem solving.

Teacher note: This is one of the most complex activities in book. No one part is that hard, but there are many parts to it. You should go through the setup and procedure on your own, making sure you are comfortable with the instructions before you have to explain it all to a demanding group of students. You will also want to make sure all the materials are obtained and prepared for use before presenting the lesson. Classroom management will be the key to success. It's useful to let students know what their responsibilities are and what your

MATERIALS (*continued*)

- 2 light meters*

- 4 2 in. Styrofoam balls

- 4 ⅜ in. × 4½ in. dowels

- 4 2 in. × 2 in. Styrofoam blocks (size is flexible as long as they are large enough to hold 2 in. Styrofoam balls on dowels without flipping over)

- Chalk, masking tape, or nonpermanent markers (for marking on the floor)

- 2 penlight-style flashlights

- ¼ in. graph paper (numerous sheets)

*   There are many sources for light meters. The easiest kind for students to use has a digital readout, giving numbers that indicate light intensity (measured in units of lux in the metric system and foot-candles in the British system.) There are relatively inexpensive models either with digital readouts or with analog meters (needle sliding along a scale) that work well and can be purchased for around $20 on the internet. You can also use calculator-based laboratories (CBLs) or other interface devices that have photometer probes. Some high school chemistry or physics teachers have these.

(continued on p. 95)

expectations are for their behavior when they are doing their independent research.

You may decide after reading through this activity that you will need to modify this lesson. You will need to judge the makeup of your class, look at the materials you can gather, and consider the constraints for having students work independently in darkened rooms. There are a few options you can consider, although none of them are as rich, engaging, and student-oriented as carrying out the lesson as written. One option would be to do the first research team activity ("Effect of Earth's Changing Distance to the Sun") as a demonstration. Then allow student groups to do the other two research activities. Another option might be to do all the activities as demonstrations and have students collect data as a group. You will need to decide what works best for you in your situation, but the more work the students do, the better.

You should also be aware that the first research team activity produces a "null" result. It shows that the slight change in distance to the Sun (a 3% variation) is not significant enough to produce the seasonal effects experienced on the Earth. But what this means is that other factors in the observations for the first activity may produce wider variation in the data collected. This may confuse students. Encourage them to think about other factors in the observations that might explain the variation (e.g., Is the light meter pointed directly at the lightbulb at each test location? Is there more light bouncing off the walls behind the bulb in one test location than another?). Null results like students experience in this situation are typical in real science, so this is an effective learning tool to help students grapple with considering extraneous factors in an experiment that could influence the data collected.

MATERIALS (continued)

- 2 copies of the "Angle Indicator Sheet" (p. 108)

- 2 manila file folders

- 4 large paperclips

- Black paper (enough to produce two 56 in. × 56 in. squares)

- Markers to write on black paper (gold or silver broad-point pens work well)

- 6 stick pins of three different colors (need all of these if optional activities are done)

- 4 safety pins that fit snugly around ⅜ in. dowel

- 4 straight pins

- 4 extension cords

- Copies of each group's assignment sheets

- Cheap black plastic tablecloths or other material for covering windows to darken the room for the groups doing Research Topic 1, "Effect of Earth's Changing Distance to the Sun." If students will be working in more than one classroom, you will need enough to darken all windows. Research Topics 2 and 3 do not need totally darkened rooms.

- Astronomy lab notebooks

Three-Dimensional Learning Exposed

Although many of the other experiences in this chapter involve three-dimensional learning, the "Reasons for the Seasons Symposium" provides more opportunity for three-dimensional thinking than any other experience in the book. It puts the student in the position of thinking and acting like scientists—from making observations, to interpreting data, to making claims from the data and then presenting the results to others in a symposium format.

Not only does the experience integrate the disciplinary core ideas, science practices, and crosscutting concepts from the *Next Generation Science Standards*, it also includes many language arts and mathematics activities that align with the *Common Core State Standards:*

- *Mathematics:* Students summarize and graph numerical data sets associated with the height of the Sun above the horizon at different times of year.

- *Speaking and listening:* Students engage effectively in a range of collaborative discussions with diverse partners, building on others' ideas and expressing their own during the final symposium and in their small-group research teams.

- *Writing:* In preparation for the symposium presentations, students need to write arguments that support claims with logical reasoning and relevant evidence. One should really consider this a six-dimensional learning experience.

As teachers discover, the "Reasons for the Seasons Symposium" is the most complex and challenging experience in this book to implement, but it is perhaps also the most representative of the kind of learning and thinking expected in the *Next Generation Science Standards,* so it is well worth the effort.

Procedures

(This has been divided into four different sections.)

PREASSESSMENT ACTIVITY

PROCEDURE

1. Revisit the discussion that the students had during Experience 2.2, "How Can This Be True?" Have students restate what they said regarding how there could be such different weather (i.e., seasons) on the Earth at the same time. The goal is not to come to any conclusions at this time but to get them thinking about how weather differs in various places on the Earth on the same date and how this might relate to the cause of the seasons. Resolving why these weather differences exist is a good motivator for understanding the seasons and conducting the activities in the rest of this experience.

2. Tell the students that to fully understand the differences in the two e-mails, they need to understand what causes the seasons on the Earth. Indicate that they will be doing research over a number of days to explore various factors that might cause Earth's seasons and whether each one can explain the weather differences in the e-mails.

3. To help the students start thinking about how the Earth and Sun cause the seasons, have the students revisit what they wrote in their astronomy lab notebooks in response to question 3 of Experience 2.1, "What Do We Think We Know?" Encourage them to make any modifications based on changes in their thinking since you did that activity. Ask them to write the explanation in their astronomy lab notebooks, including diagrams that will help the reader understand what they are saying. If you did not do Experience 2.1 or 2.2, an effective way to get at students' preconceptions regarding what causes the seasons is to have them pretend they are writing a picture book for students two years younger than them that explains Earth's seasons.

MATERIALS

For the class:

- Space on a whiteboard, blackboard, or piece of poster paper

One per student:

- Astronomy lab notebook

4. Ask various individuals to present their explanations and encourage discussion of the various ideas. Let students know that there is more than one factor that contributes to Earth having seasons.

5. Conclude the discussion by summarizing on poster paper or the blackboard all the factors they suggest might cause the seasons. The discussion will most likely include the following ideas:

    • The changing distance of the Earth from the Sun

    • The Sun's height above the horizon during the day

    • The varying number of hours of sunlight on different days of the year

    • Tilt of Earth's axis

    • Orbit (revolution) of the Earth around the Sun

    • Spin (rotation) of the Earth on its axis

    • Local geography

    • Weather and clouds

If the first four factors have not been suggested, use the discussion to make sure they are included.

SYMPOSIUM INTRODUCTION AND RESEARCH TOPIC ASSIGNMENT

PROCEDURE

MATERIALS

See items listed under each research topic.

1. Tell the students that these days scientists typically work in teams of researchers at a university, research center, or both. Once or twice a year, these teams will meet with other research teams at conferences or symposia to share and compare their observations. Tell the students that for the next few sessions they will work in teams to conduct research on one possible cause of the seasons and then share their results with other groups.

2. Organize the class into research teams of four to five students. The teams will conduct experiments to understand the effect of the first six factors identified in the previous step. Since several of these factors are interrelated, some of the teams will examine more than one factor. Ideally, there should be six teams so that two teams are conducting the same research. This will allow for an appropriate comparison of results and for more effective discussion.

3. Provide each group with the equipment and assignment sheets for conducting their research, as noted in each research activity that follows. The introduction of the research topics is a critical point in this problem-solving approach to studying the seasons. Some students may feel that they do not have all the information they need, especially if they are used to a more directed approach to studying science. Some will have difficulty understanding the Sun and Earth model used in the activities. Assure them that this feeling is reasonable and expected and is much like the nervousness many scientists feel when they explore a new subject. Tell them to discuss questions they have with their team members to see if the group can resolve them without assistance. If they still have lingering questions, ask them to write these down in their astronomy lab notebooks and begin doing the experiments to the best of their abilities.

4. Give the students some time to begin doing the experiments, and then move among the groups to answer questions and help them understand how the equipment and process works. Students should keep a log of activities, questions, and observations during their research.

RESEARCH TOPIC 1: EFFECT OF EARTH'S CHANGING DISTANCE TO THE SUN

OVERALL CONCEPT

In this research activity, the team makes a scale model of the Earth orbiting the Sun and investigates the amount of light falling on the Earth at various times during its yearly orbit. They will come to understand how

Earth's orbit is almost a circle and thus cannot be a factor in what causes Earth's seasons.

ADVANCE PREPARATION

You will need to identify a space approximately 15 ft. × 15 ft. in size that can be made dark for this activity. It can be done in a smaller space, but the larger the area, the better the model works. The activity also works best if the model is in the center of the room, equidistant from the walls. Reflected light off the walls can add errors to the light measurements, which can confuse students. At the same time, this can lead to some great discussion about errors in measurements and the potential causes.

You will also need to locate a light meter. The best source for this is a friend or perhaps a photography teacher—otherwise you can purchase an inexpensive one online or at your local camera store. See the note on light meters in the materials list for the whole lesson (p. 94). When students start using the light meter, you will want to demonstrate its use and make note of the units measured by it. The absolute intensity of the light level measured is not critical and will vary depending on the exact size of the model orbit and the color of the walls. What is critical is the difference in the measurements made at each location in the orbit.

Next, you will need to assemble the model Earth setup. Obtain a block of Styrofoam that is approximately a 2 in. cube. Make a copy of the timing circle in Figure 2.6. Cut out the timing circle and the hole in the middle of the circle. Push the Styrofoam ball onto the dowel about 1 in. Place a small safety pin around the shaft of the dowel. Push the dowel through the middle of the timing circle and into the Styrofoam cube at about a 23° angle from the perpendicular. Push the timing circle with the safety pin on top of it into the top of the cube. Place one straight pin through the circle into the cube to hold the circle in place while the stick is rotated. The assembled items will look like Figure 2.7.

MATERIALS

For the class:

- Model Earth set up as shown in Figure 2.7 (2 in. Styrofoam ball on ⅜ in. × 4 ½ in. dowel placed in 2 in. Styrofoam cube at 23° angle from perpendicular, one straight pin)

One per group:

- Room that can be made completely dark

- Short lamp with 250 W lightbulb, shade removed

- Measuring tape (preferred) or meterstick

- Chalk, masking tape, or a nonpermanent marker (to mark the floor)

- Light meter (see note on p. 94)

- "Reasons for the Seasons Symposium: Earth's Changing Distance Research Team Assignment Sheet" (pp. 102–103)

- Timing circle (cut from template, Figure 2.6)

- Room that can be made completely dark

Place all equipment for each research team in an appropriate container (resealable plastic bags work well) and make a copy of the "Reasons for the Seasons Symposium: Earth's Changing Distance Research Team Assignment Sheet" for each team member.

PROCEDURE

1. Give each team a container of equipment and identify the team's research space.

2. Have students read the research instructions, discuss among themselves what they are to do, and list any questions they have in their astronomy lab notebooks.

3. When they are done, have the students review their research plans with you so you can clarify any issues and answer any questions. If the operation of the light meter is not obvious, show them how to use it.

4. Let the research groups proceed as independently as possible to work through the activity using the instructions on the assignment sheet.

FIGURE 2.6

Timing circle

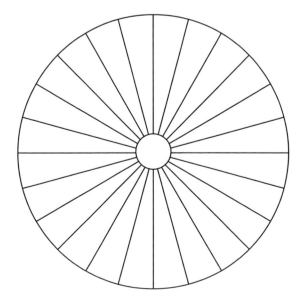

FIGURE 2.7

Assembled timing circle apparatus

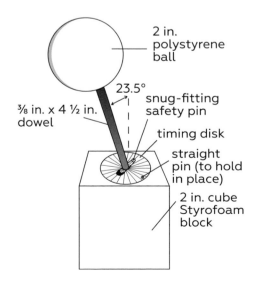

2 in. polystyrene ball

23.5°

snug-fitting safety pin

³⁄₈ in. x 4 ½ in. dowel

timing disk

straight pin (to hold in place)

2 in. cube Styrofoam block

Reasons for the Seasons Symposium: Earth's Changing Distance to the Sun

RESEARCH TEAM ASSIGNMENT SHEET

Research Goal

Your goal is to determine what effect the changing distance of the Earth from the Sun has on Earth's seasons.

Background

You will be making a scale model of Earth's orbit around the Sun and measuring how much the intensity of the Sun's light changes from the time of Earth's closest approach to the Sun in early January to when Earth is farthest from the Sun in July. Be sure to keep a log in your astronomy lab notebook. Note the steps you follow throughout the activity and include your results and conclusions.

Procedure

1. Determine the distance from the Sun to the Earth at different times of year in your scale model using the scale of 2,000,000 mi. = 1 in. Enter your calculations in the chart below. The scale can be adjusted if necessary to make the model fit better in your research space.

2. Use the information from step 1 and the diagram on the next page to construct a scale model of Earth orbiting the Sun. Use the lamp as the Sun and the Styrofoam ball and holder as the Earth. This experiment works best if the model is in the center of

MATERIALS

- 250 W lightbulb in a short desk lamp, shade removed (*Safety note:* This lightbulb will get very hot. Be careful not to touch it.)

- Measuring tape or meterstick

- Light meter

- Chalk, masking tape, or nonpermanent marker to mark the floor

- Model Earth provided by your teacher (Styrofoam ball on stick placed in Styrofoam cube at an angle of 23.5°)

- "Reasons for the Seasons Symposium: Earth's Changing Distance to the Sun Research Team Assignment Sheet"

- Room that can be made completely dark

Date	Distance from Sun to Earth (mi.)	Distance in scale model (in.)
July 4	94,550,000 (Earth farthest from Sun)	
Oct. 4	92,970,000	
Jan. 4	91,390,000 (Earth closest to Sun)	
April 4	92,970,000	

Reasons for the Seasons Symposium: Earth's Changing Distance to the Sun

RESEARCH TEAM ASSIGNMENT SHEET

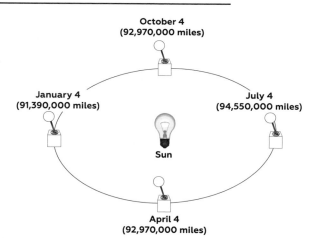

October 4
(92,970,000 miles)

January 4
(91,390,000 miles)

July 4
(94,550,000 miles)

Sun

April 4
(92,970,000 miles)

the room so that the walls of the room are at the same distance from the lamp. Mark on the floor where Earth is located for each of the dates listed. Also mark on the floor which direction north is in your model—in approximately the same direction as the January 4 location in your model.

3. Now that your model is built, place the model Earth in each of the four positions and use the light meter to measure the brightness of the Sun from that location. Be sure the model's North Pole is always pointing in the north direction when you place your model Earth at each location, and be sure the light meter is pointing directly at the model Sun when you take the reading. Record the results in your astronomy lab notebook. Be sure to go around to the four locations several times and average the results. If you find differences between the measurements, discuss with your teammates what might cause these differences.

4. To get a sense of how great an effect Earth's changing distance from the Sun really has on the seasons, it is useful to look at the percent change in Earth's distance when it is farthest from the Sun versus when it is closest the Sun. This percent change is

$$\frac{\textbf{(farthest distance to the Sun − closest distance to the Sun)}}{\textbf{farthest distance from the Sun}} \times \textbf{100} = \textbf{\% (percentage) change}$$

Include this information in your presentation to the class.

5. Using the results from your experiment, have a discussion among members of your research team to predict what effect the changing distance of Earth from the Sun has on Earth's seasons. Write your conclusions in your astronomy lab notebook.

6. Prepare a presentation for the rest of the class, using appropriate visual aids, that explains the research activity you performed and what conclusions you reached. Be sure to include the following:

    •   The problem you explored

    •   The procedure you followed

    •   The data you collected, including any graphs

    •   The conclusions you reached

RESEARCH TOPIC 2: EFFECT OF THE SUN'S HEIGHT ABOVE THE HORIZON

OVERALL CONCEPT

In this activity, the research team explores the effect of the Sun's height above the horizon on the amount of energy received by a given location on Earth. They then infer what influence this might have on Earth's seasons.

ADVANCE PREPARATION

Place all equipment (except the books) for the research team in an appropriate container (resealable plastic bags work well). Make a copy of the assignment sheet for each team member.

PROCEDURE

1. Give each team a container of equipment and identify the team's research space.

2. Have students read the research instructions, discuss among themselves what they are to do, and list any questions they have in their astronomy lab notebooks.

3. When they are done, have them review their research plans with you so you can clarify any issues and answer any of their questions.

4. Let the research group proceed as independently as possible as they work through the activity using the instructions on the "Reasons for the Seasons Symposium: Sun's Height Above the Horizon Research Team Assignment Sheet."

MATERIALS

For the class:

- Room that is relatively dark with lights off

One per group:

- Penlight-style flashlight

- ¼ in. × ¼ in. graph paper

- "Angle Indicator Sheet" with Height of Sun at Noon Chart (part of the assignment sheet, p. 108)

- Manila folder

- Two large paper clips

- "Reasons for the Seasons Symposium: Sun's Height Above the Horizon Research Team Assignment Sheet" (p. 105–108)

- Tape or glue

- Several thick books

Reasons for the Seasons Symposium: Sun's Height Above the Horizon
RESEARCH TEAM ASSIGNMENT SHEET

Research Goal

Your assignment is to determine what effect the changing height of the Sun above the horizon will have on the Earth's seasons.

Background

As you may have noticed, each morning the Sun is near the horizon when it rises and then climbs higher in the sky. It reaches its highest point above the horizon around noon and then gets lower in the sky until it is again near on the horizon at sunset. The highest point that the Sun reaches above the horizon around noon changes throughout the year. In the United States and other countries at the same latitude, the Sun is highest in the sky around noon on a date astronomers call the summer solstice (June 20–22). The Sun is lowest in the sky around noon on the winter solstice (December 20–22).

In this research activity, you will simulate the Sun's changing height above the horizon at noon using a flashlight to represent the Sun. You will then use the data you collect to determine what effect this may have on Earth's seasons. Be sure to keep a log in your astronomy lab notebook of the steps you follow throughout the activity and include any results and conclusions.

MATERIALS

- Penlight-style flashlight

- ¼ in. × ¼ in. graph paper

- "Angle Indicator Sheet" with Height of Sun at Noon Chart

- Manila folder

- 2 large paper clips

- Space to work that is relatively dark

- "Reasons for the Seasons Symposium: Sun's Height Above the Horizon Research Team Assignment Sheet"

- Tape or glue

- Several thick books

Procedure

1. Find a table in a darkened space where you can lay out your materials.

2. Open the manila folder and paper clip the graph paper to the inside of the right half of the folder, as shown in the diagram (see next page).

3. Tape or glue the "Angle Indicator Sheet" to the inside of the left half of your folder as shown in the diagram. The sheet should be folded at the bottom so that the 0° line is on the fold of the manila folder.

4. Set up the folder so that the right half is flat on the table and the left half is at a 90° angle, as shown in the diagram. Either someone can hold the folder open so it forms a 90° angle or a few stacked books can be used to hold the left side in place—a small piece of tape can help to keep the left side of the folder next to the books.

Reasons for the Seasons Symposium: Sun's Height Above the Horizon
RESEARCH TEAM ASSIGNMENT SHEET

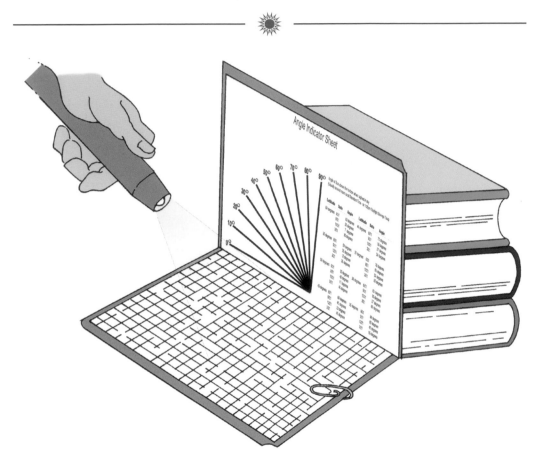

5. Identify the latitude of your location by looking on a world map or searching online. Then find the closest latitude listed on the Angle of the Sun Above the Horizon Chart on the "Angle Indicator Sheet." This chart gives the height of the Sun above the horizon for your latitude at different times of the year in the Northern Hemisphere.

6. To simulate the Sun shining on the Earth at noon, turn on the flashlight and place it at the end of the angle line for the height of the Sun above the horizon at noon on the summer solstice, as shown in the diagram above. You will need to develop a way to be sure the flashlight is always

   a. held pointing straight along the angle line; and

   b. the same distance out from the "Angle Indicator Sheet," so that the entire flashlight beam falls on the graph paper.

   Draw an outline of the flashlight beam on the graph paper.

Reasons for the Seasons Symposium: Sun's Height Above the Horizon
RESEARCH TEAM ASSIGNMENT SHEET

7. Remove the graph paper, being sure to record what date the outline represents (e.g., summer solstice).

8. Have one team member count the number of squares covered by the flashlight beam. Be sure to develop a method to account for partially covered squares. Record the results in your astronomy lab notebook. Repeat the count of squares several times and get an average.

9. Repeat the process for other times of the year, and produce a graph in your astronomy lab notebook that represents the area covered by the flashlight beam at different times of the year. (Remember which things you need to keep the same, as explained in step 6.)

10. (*Optional*) If you have time, record the percentage change in the area covered by the flashlight beam between the first date and each of the rest of the dates. Record this in your astronomy lab notebook. This can be calculated using the following formula:

$$\frac{\textbf{(area covered by light on first date - area covered by light on second date)}}{\textbf{area covered by light on first date}} \times 100 = \% \text{ (percentage) change}$$

11. Remembering that the flashlight represents the Sun, have a discussion among members of your group to predict what effect the changing height of the Sun above the horizon at noon will have on Earth's seasons.

12. Using appropriate visual aids, prepare a presentation for the rest of the class that explains the research activity you performed and what conclusions you reached. Be sure to include the following:

    • The problem you explored

    • The procedure you followed

    • The data you collected, including any graphs

    • The conclusions you reached

2.5

Angle Indicator Sheet

Angle of Sun above the horizon when highest in sky
(Usually around noon local Standard Time - or 1:00pm Daylight Savings Time)

Latitude	Date	Angle	Latitude	Date	Angle
60 degrees	6/21	53 degrees	40 degrees	6/21	73 degrees
	9/21	30 degrees		9/21	50 degrees
	12/21	7 degrees		12/21	27 degrees
	3/21	30 degrees		3/21	50 degrees
55 degrees	6/21	58 degrees	35 degrees	6/21	78 degrees
	9/21	35 degrees		9/21	55 degrees
	12/21	12 degrees		12/21	32 degrees
	3/21	35 degrees		3/21	55 degrees
50 degrees	6/21	63 degrees	30 degrees	6/21	83 degrees
	9/21	40 degrees		9/21	60 degrees
	12/21	17 degrees		12/21	37 degrees
	3/21	40 degrees		3/21	60 degrees
45 degrees	6/21	68 degrees	25 degrees	6/21	88 degrees
	9/21	45 degrees		9/21	65 degrees
	12/21	22 degrees		12/21	42 degrees
	3/21	45 degrees		3/21	65 degrees

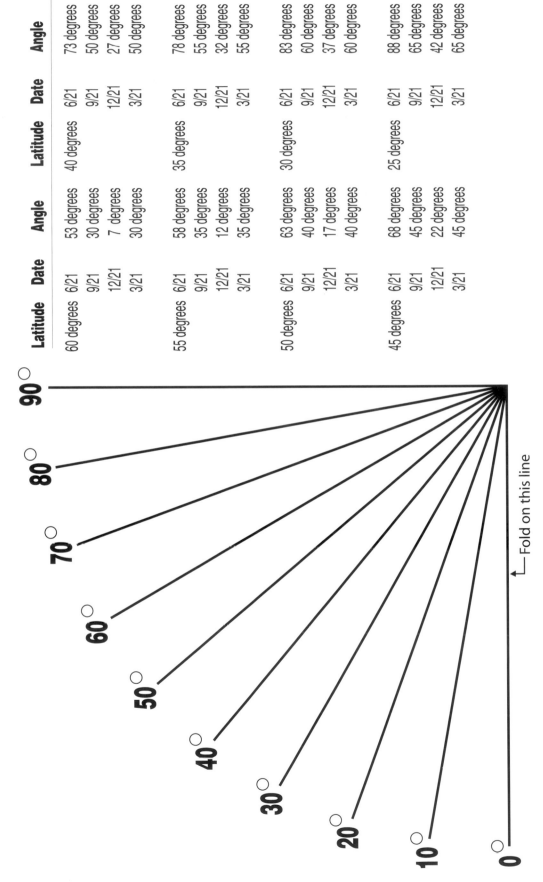

Fold on this line

90
80
70
60
50
40
30
20
10
0

RESEARCH TOPIC 3: EFFECT OF THE DIRECTION OF EARTH'S AXIS ON THE SEASONS

OVERALL CONCEPT

In this activity, the research team makes a scale model of the Earth orbiting the Sun to study how the direction of Earth's axis of rotation affects the number of hours of daylight that various parts of Earth experience during different seasons. They then predict what effect this might have on Earth's seasons.

ADVANCE PREPARATION

The first time you do this activity, you will need to copy the drawing of Earth's orbit shown in Figure 2.8 (p. 110) onto pieces of black butcher paper taped together on the back to make an area approximately 56 in. × 56 in. Most art stores have rolls of black paper and silver or gold broad-point rollerball pens that work well to reproduce the drawing. After the first use, store the drawing with the other materials for later. Consider laminating the drawing if possible. You will also need to assemble the model Earth setup (see Figures 2.6 and 2.7 and detailed assembly instructions on p. 101). The Styrofoam ball and stick are easily found in craft stores.

Be sure to use a short enough lamp or place the model Earth on a stand so that the lamp and Earth are at the same level when in use. Identify a workspace that can be somewhat darkened. Place all equipment for each research team in an appropriate container and make a copy of the assignment sheet for each team member.

PROCEDURE

1. Give each team a container of equipment and identify the team's research space.

2. Have students read the research instructions, discuss among themselves what they are to do, and list any questions they have in their astronomy lab notebooks.

MATERIALS

One per group:

- 40 W lightbulb in a short desk lamp, shade removed

- Black butcher paper showing orbit of Earth around Sun

- Model Earth set up as shown in Figure 2.7 on page 101 (2 in. Styrofoam ball on ⅜ in. × 4 ½ in. stick placed in 2 in. Styrofoam cube at 23° angle from perpendicular; timing disk; one straight pin; one safety pin)

- Three map pins of different colors

- "Reasons for the Seasons Symposium: Direction of Earth's Axis Research Team Assignment Sheet" (pp. 111–113)

3. When they are done, have them review their research plans with you so you can clarify any issues and answer any of their questions.

4. Let each research group proceed as independently as possible to work through the activity using the instructions on the "Direction of Earth's Axis Research Team Assignment Sheet."

FIGURE 2.8

Drawing of Earth's orbit on butcher paper

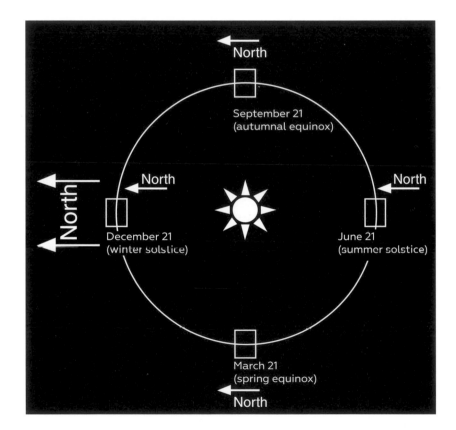

Reasons for the Seasons Symposium: Direction of Earth's Axis
RESEARCH TEAM ASSIGNMENT SHEET

Research Goal

Your goal is to determine what effect the direction of Earth's axis of rotation has on the number of hours of daylight received at different locations on Earth and how this might affect Earth's seasons.

Background

As you may already know, Earth spins on its axis every 24 hours, as if someone pushed a large needle through Earth from the North Pole to the South Pole and then made Earth spin on the needle. The direction of Earth's axis is not perpendicular to the direction to the Sun but tilted by 23.5°. In this activity, you will make a model that shows the Sun and the Earth and how the Earth orbits the Sun. This will let you determine how much daylight shines on different parts of Earth at different times of the year. You will then use this information to predict what effect this may have on Earth's seasons.

Be sure to keep a log in your astronomy lab notebook of the steps you follow throughout the activity and include any results and conclusions you make.

Procedure

1. Find a darkened space where you can carry out your research.

2. Lay out the black paper showing Earth's orbit. Be sure to leave room around all sides for your team to work.

3. Place the lamp with the lightbulb in the Sun's location on the black paper, and turn on the lamp.

4. Be sure the model Earth setup is assembled as shown in the diagram to the right. Be sure the safety pin fits snugly on the wood stick. Also, make sure the model Earth and the lightbulb are at the same height.

MATERIALS

- 40 W lightbulb in short desk lamp, shade removed (*Safety note:* This lightbulb will get very hot. Be careful not to touch it.)

- Black butcher paper showing orbit of Earth around Sun

- Model Earth setup (Styrofoam ball on dowel, 2 in. × 2 in. Styrofoam block, timing circle, and a safety pin that just fits around the dowel; see diagram below)

- Three map pins of different colors

- "Reasons for the Seasons Symposium: Direction of Earth's Axis Research Team Assignment Sheet"

- Space to work that is relatively dark

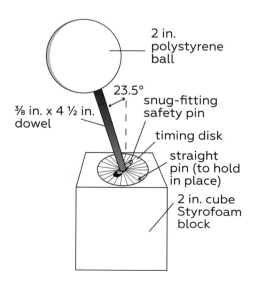

Reasons for the Seasons Symposium: Direction of Earth's Axis
RESEARCH TEAM ASSIGNMENT SHEET

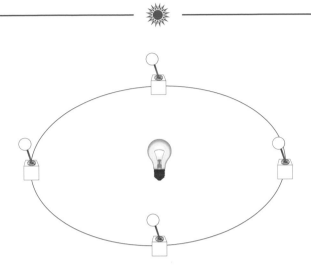

5. Find the latitude of your location on a globe or by searching online and place one of the map pins near your latitude on your model Earth. Push the pin in so the head is close to the surface of the Styrofoam ball.

6. Set the model Earth on the poster paper at the location marked summer solstice. Be sure Earth's axis (the stick) is tilted with the top of the stick pointing toward the direction north shown on the poster paper (as shown in the left side of the diagram on p. 110).

7. Earth rotates counterclockwise on its axis (when viewed from above) once every 24 hours. Practice turning the model Earth in the proper direction and note that the safety pin also makes one complete turn and passes across all 24 hour indicators on the timing circle with each complete rotation.

8. You are now ready to use your model Sun and Earth to determine the number of hours of daylight at your location on Earth at different times of the year.

9. Turn Earth counterclockwise until the map pin marking your location is just moving from the nighttime side (dark side) of your model Earth into the daylight side (lighted side). This is when the Sun would rise for a model person riding along with the pin. Hold the model Earth in place while another team member moves the safety pin so it is over one of the hour lines.

10. Now rotate Earth until the map pin is at sunset (where the pin goes from the lit side of the model Earth into the dark side). On the timing circle, count the number of hours the safety pin has passed over (including an estimate of any fraction of an hour at the time of sunset). This is the number of hours of sunlight you experience at your location on the summer solstice.

Page 3

Reasons for the Seasons Symposium: Direction of Earth's Axis
RESEARCH TEAM ASSIGNMENT SHEET

11. Record this information in your astronomy lab notebook. Let everyone in the group do this so that you repeat the observation for the summer solstice several times and record the average number of hours of sunlight your group measured.

12. Move the model Earth to each of the other three locations marked on the poster paper and repeat the observations. *Be sure that Earth's axis is always pointed toward the north direction at each location.* Record the data in your astronomy lab notebook. Then prepare a table and a graph of the data you collected that shows how the amount of daylight varies at different times of year.

13. Use the data to have a discussion among the members of your research group to predict what effect the tilt of Earth's axis will have on Earth's seasons.

14. Using appropriate visual aids, prepare a presentation for the rest of the class that explains the research activity you performed and what conclusions you reached. Be sure to include the following:

   • The problem you explored

   • The procedure you followed

   • The data you collected, including any graphs

   • The conclusions you reached

15. (*Optional*) Repeat your observations, except remove the tilt of Earth's axis toward or away from the Sun. What can you conclude about how daylight hours would change if the Earth's axis were not tilted?

16. (*Optional*) Tilt the Earth again and place a new map pin at the equator. Repeat your observations at all four positions in Earth's orbit around the Sun. Record these observations in your astronomy lab notebook. Discuss with your group what you think the implication for the seasons for people who live near the equator is. Record your predictions in your astronomy lab notebook.

17. (*Optional*) With the Earth still tilted, place a new map pin at a location on the other side of the equator (in Earth's Southern Hemisphere) that is the same distance away from the equator as your home location. Repeat your observations at all four positions (dates) in Earth's orbit around the Sun. Record these observations in your astronomy lab notebook. Discuss with your research team what these observations tell you about the length of daylight at different locations on the Earth and what this means in terms of the Earth's seasons.

HOLDING THE SCIENTIFIC SYMPOSIUM

Just as scientists meet regularly to share their latest observations, data, and discoveries, this part of the activity allows students to share the results of their research activities with their peers. Each team should have a presentation prepared that describes the following:

- The problem they explored

- The procedure they followed

- The data they collected, including any graphs

- The conclusions they reached

PROCEDURE

1. Have each group present its findings to the rest of the class. If two groups are doing the same research project, they should each have an opportunity to present. After each presentation, the poster papers prepared by each group should be left up around the room for use in later discussions.

2. It works best to leave questions and discussion until after all groups have presented their results. Have each student take notes in his or her astronomy lab notebook about each group's presentation and write down any questions they have, which can be asked later during the whole-class discussion.

3. Once every group has presented, let students ask any questions they have. Encourage as much student-to-student discussion as possible without your taking a leadership role in the interaction. The discussion will usually lead to identifying which factors affect Earth's seasons. If not, ask the question of the class after the discussion of results is well underway.

4. At the end of the all-class discussion, summarize the conclusions you reached, which will usually include the following:

   a. The yearly change in the distance of Earth to the Sun is such a small percentage of the distance (and the resulting change in sunlight received is so small) that it does not have an influence on Earth's seasons—and in the

MATERIALS

- Presentation materials produced by each team

- Whiteboard space to write down the main conclusion from the research

Northern Hemisphere, we are actually closer to the Sun in winter than in summer.

b. The tilt of Earth's axis causes Earth to have longer days in the summer and shorter days in the winter. A longer day means the Sun has more hours to heat the Earth's surface.

c. The higher the Sun gets in the sky at noon, the more concentrated the sunlight is that falls on that particular location on Earth. The Sun gets higher during the summer. When light is more concentrated, it is more effective at heating, so the parts of the Earth under a higher Sun get heated more effectively.

5. Conclude by reinforcing the simple concept that it is the tilt of Earth's axis that causes both the longer day and more concentrated sunlight during the summer—and vice versa for the winter.

6. As a postassessment activity, have students write a revised version of their children's activity book. See instructions in Experience 2.9 (p. 139) in the Evaluate section of this chapter.

EXPERIENCE 2.6

Length of Day Around the World

Overall Concept

This activity builds from Experience 1.5, "Noontime Around the World." Students use an Earth globe and lamp to model the Sun–Earth system. Gnomons are attached to the globe at one location in the Southern Hemisphere, one in the Northern Hemisphere, and one on the equator (all at the same longitude). Other gnomons can be added as desired, such as for the poles or locations above the Arctic Circle. Students have their "Earth" orbit the "Sun," making sure the Earth's axis always points in the same direction in space. Students observe how the amount of daylight and the height of Sun above the horizon (as measured by using the length of each gnomon's shadow) vary depending on the gnomon's latitude and the season. A full page of gnomon recording circle templates is available at *www.nsta.org/solarscience*.

Objectives

Students will understand that

1. noon occurs at the same time for places on a given longitude;

2. the gnomon's shadow will always fall along the north–south longitude line at noon, but the length of the shadow will vary depending on the location's latitude;

3. the amount of daylight that a given location receives depends on its latitude; and

4. when it is noon in their location, most other places on the Earth are experiencing a different time of day.

MATERIALS

Per group of three to four students:

- Clamp lamp with directional shade that shines the light in one direction

- Earth globe

- Several small balls of clay or picture-hanging putty

- Several ends of toothpicks (cut to 8 mm long)

- Several gnomon shadow recording circles (see template in Figure 1.10, p. 37)

One per student:

- "Length of Day around the World Discussion Questions" (p. 123)

- Astronomy lab notebook

Understanding and Tracking the Annual Motion of the Sun and the Seasons

Advance Preparation

1. Make sure the room can be made relatively dark—although it does not need to be as dark as for Experience 1.4, "Modeling the Sun–Earth Relationship," in Chapter 1.

2. Be sure that students have enough room and table space to set up their Sun–Earth systems, so that their model Earths and Suns are at least four feet apart. They also need room for the model Earth to orbit the model Sun.

3. Identify a direction in the room that will be north and place a sign on the wall as close to the ceiling as you can reach. Ideally, this north is the same as magnetic north, but it can be in any direction for the purpose of this activity.

4. You will note that the activities call for students to repeat their observations four different times (summer, fall, winter, and spring). If you do not have time or room for each group to make all four observations, you can consider having each group do one or two of the locations and then share their results.

Procedure

1. Have students build three model sundials from balls of putty that are about the size of a small pea, gnomon shadow recording circles, and toothpicks as shown in Figure 2.9.

FIGURE 2.9

Diagram showing how to build a model sundial

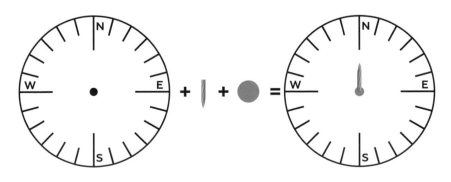

Cut out enough "Gnomon Shadow Recording Circles" (p. 37) so that each group can build three sundials. Put a small hole in the center so that the toothpick base can stick through the hole. The toothpick sticks up through the hole with the putty underneath, which will hold it in place and attach it to the globe.

FIGURE 2.10

A model Sun and Earth setup

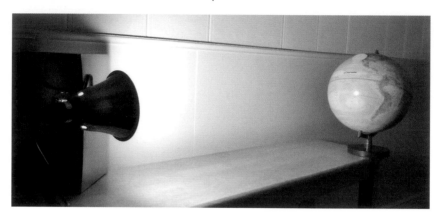

Note that the model Sun is pointed directly at the globe's equator.

2. Have students set up their globe and lamp with the lamp's shade directing the Sun's light on the Earth and with the lamp at the same height above the table as the Earth's equator (Figure 2.10). The axis of rotation of the globe should be pointed away from the direction of the Sun (i.e., with the North Pole tipped away from the direction of the Sun), which represents the Earth's orientation relative to the Sun on the winter solstice (the first day of winter in the Northern Hemisphere, which occurs around December 21). The globe also needs to be positioned in its orbit around the Sun so that the North Pole is pointing in the north direction in the room as indicated by the sign on the wall.

3. Have the students attach one sundial to the globe at their location. Have them attach a second at the equator, and a third about the same distance south of the equator as their home location. All three should be on the same longitude line. Be sure the north–south lines on the sundials are lined up to point to the North and South Poles of the globe (Figure 2.11).

4. Assuming students did Experience 1.5, "Noontime Around the World," from Chapter 1, you may wish to skip this step. If they have not worked with the globes and model sundials before, have students explore their model Earth–Sun system to answer the following questions in their astronomy lab notebooks (be

sure they always turn the globe in a counterclockwise direction when looking down on the North Pole):

a. Turn your globe so that it is sunrise for your model sundial. Where is the sundial at your location relative to the Sun when it is sunrise?

b. Where is the sundial when it is sunset at your location?

c. Where is the sundial when it is noon at your location?

d. Which part of the United States experiences sunset first? Which part experiences it last?

These four questions give you time to observe the groups to see how much they understand and who needs extra help to work with the model. After students have had time to explore using the model Sun–Earth system, have a whole-class discussion of what the students discovered, reinforcing that the Sun rises along the eastern horizon and sets along the western horizon. Plus, the sundial is facing toward the Sun at noon with the shadow falling along the north–south line.

Teacher note: Depending on what globe you use, the support arm holding the globe in place may not allow the gnomon to clear the arm as the globe is turned. Students will need to adjust the gnomon to stick straight out after the sundial passes under the support arm. Also, this experience can be a challenge for students because

FIGURE 2.11

Three sundials along same longitude line

Note that north–south lines of the sundials are lined up with north and south on the globe.

of the spatial reasoning required. It will help if they think of themselves as miniature people standing next to the sundial as they work to answer these questions. Finally, students may have a hard time knowing exactly when sunrise and sunset occur on their globes. The best indicator of when a location on the globe has moved from daytime to nighttime is when the students can no longer see a shadow from the gnomon. Sunrise is when the students can first see the gnomon's shadow.

5. Just before starting the students on their group work, point out the direction for north in the room and remind them that as they have their model Earths orbit the Sun, they will always need the Earth's North Pole pointing in that direction. It is best to demonstrate this with one of the Sun–Earth system models (i.e., walk the globe around the Sun with the North Pole always pointing in the north direction).

6. Have them work in their groups to complete the "Length of Day Around the World" worksheet. Have a whole-class discussion after they complete their group discussion and have students attach their answers in their astronomy lab notebooks.

Length of Day Around the World
Discussion Questions
TEACHER VERSION*

1. Start at the point in the Earth's orbit around the Sun where the Earth's axis is tipped away from the Sun (the winter solstice orientation, around December 21). What happens to the shadows on your three model sundials as you rotate the globe to take your location from sunrise to noon to sunset? Include answers to the following questions in your observations:

    a. At sunrise, noon, and sunset for your location, where are the other sundials relative to sunrise, noon, and sunset at their locations?

    b. How does the length and direction of the gnomon shadows differ for each location and time?

    c. Which location has the greatest amount of daylight? Which has the least?

    At our sunrise, the other two locations have already experienced sunrise. At noon, the shadows all fall along the north–south line, with the shadow being longest for the location in the Northern Hemisphere. Our location has the least amount of daylight, then a person on the equator, while a person in the Southern Hemisphere has the most amount of daylight.

2. Have your model Earth "orbit" the Sun for three months, which will move the Earth to the spring equinox (around March 21). Be sure that the Earth's North Pole is always pointing in the same direction in the room as you move the Earth around. Answer the three same questions given in item 1.

    All three locations experience sunrise at the same time. At noon, the shadows all fall along the north–south line, with the shadow being shortest (really no shadow) at the equator. The other two shadows are the same length but point in opposite directions (toward the north in the Northern Hemisphere and toward the south in the Southern Hemisphere). All three locations have the same amount of daylight.

3. Have your model Earth orbit for another three months to the winter solstice position in its orbit (around June 21). Again, answer the three questions given in item 1.

    The situation is the opposite of what we experienced on the winter solstice. At our sunrise, the other two locations have not experienced sunrise. At noon, the shadows all fall along the north–south line, with the shadow the longest for the location in the Southern Hemisphere. Our location has the most amount of daylight followed by a person on the equator, while a person in the Southern Hemisphere has the least amount of daylight.

Contains key information that could be included in responses. For a copy of the student handout with more space for answers, please visit www.nsta.org/solarscience.

Length of Day Around the World Discussion Questions
TEACHER VERSION

4. Have your model Earth orbit another three months to the spring equinox position (around September 21) and answer the questions again.
   The results are the same as for the fall equinox.

5. What general statement can you make regarding the seasons for different locations on the Earth?
   The seasons (and the length of the day during each season) are not the same everywhere on Earth. People in the Southern Hemisphere experience the opposite amount of sunlight in the winter and summer as people in the Northern Hemisphere. People at the equator experience the same amount of sunlight every day of the year.

Length of Day Around the World
Discussion Questions

1. Start with the Earth's axis tipped toward the Sun (the summer solstice orientation, around June 21). What happens to the shadows on your three model sundials as you rotate the globe to take your location from sunrise to noon to sunset? Include answers to the following questions in your observations:

    a. At sunrise, noon, and sunset for your location, where are the other sundials relative to sunrise, noon, and sunset at their locations?

    b. How does the length and direction of the gnomon shadows differ for each location and time?

    c. Which location has the greatest amount of daylight? Which has the least?

2. Have your model Earth "orbit" the Sun for three months, which will move the Earth to the fall equinox (around September 21). Be sure that the Earth's North Pole is always pointing in the same direction in the room as you move the Earth around. Answer the three same questions given in item 1.

Length of Day Around the World Discussion Questions

3. Have your model Earth orbit for another three months to the winter solstice position in its orbit (around December 21). Again, answer the three questions as given in item 1.

4. Have your model Earth orbit another three months to the spring equinox position (around March 21) and answer the questions again.

5. What general statement can you make regarding the seasons for different locations on the Earth?

EXPERIENCE 2.7

Seasons on Other Planets

Overall Concept

Students now apply what they have learned about seasons on Earth to the other planets in our solar system. They first learn that on other planets both the varying distance from the Sun (eccentricity of orbit) *and* the tilt of the planet's axis can influence the seasons. Then they look at characteristics of a few other planets and try to predict what the seasons on them will be like.

Objectives

Students will

1. understand that both the tilt of a planet's axis and the eccentricity of its orbit can contribute to its seasons and

2. be able to discuss what the seasons are like on various planets in our solar system.

Advance Preparation

This experience might be a bit difficult for some classes, and they may require more guidance and explanation from the teacher. It might be best to do this activity after students have studied the other planets in our solar system. It's probably best for you to read the information in the table below about the planets before you teach this experience so that you are clear on what happens on each planet you plan to cover.

> *Teacher note:* You may want to review the part about seasons on other planets in the Content Background section (p. 76). Recall that the Earth has seasons because one or more cosmic accidents tilted its axis compared to the plane of its orbit around the Sun. The Earth's

MATERIALS

Per group:

- 1 "Characteristics of the Planets" sheet (p. 129)

- 1 ball (could be an Earth globe, a tennis ball, or a somewhat larger ball)

- 2 small straws per ball to be the axes for the ball

- 2 small balls of clay or picture-hanging putty

One per student:

- Astronomy lab notebook

orbit is nearly circular, so the distance from the Sun has almost nothing to do with the seasons. The tilt dominates. For other planets, we have to ask the following:

- Is the axis tilted from the vertical? (The greater the tilt, the more pronounced the seasons.)

- How circular is the orbit around the Sun? (If it's not circular, then at some points the planet will be farther from the Sun and a bit colder and at other points it will be closer and warmer.)

Procedure

1. Review the reasons for the seasons on Earth with the class and then ask them to open their astronomy lab notebooks and write for a few minutes about the following question: "Do you think the other planets in our solar system, such as Mars or Jupiter, have seasons? Explain all the reasons why you answered yes or no in as much detail as you can."

2. Have students get together in small groups to discuss their answers and see if they can reach consensus. Have the groups report out to the whole class.

3. If the students did not come up with the two reasons that other planets can have seasons, guide them with questions to come up with both (see the Teacher Note under Advance Preparation above). Be sure they understand that other planets may have one or both these reasons for their seasons.

4. Explain the concept of the eccentricity of an orbit (see the part about seasons on other planets in this chapter's Content Background, p. 76) to give them that term as part of their science vocabulary. You may want to mention that the ancient Greeks thought of the circle as the perfect shape. Thus, it made sense that anything that is off from being a perfect circle is eccentric, or odd, in this view.

5. Give each group a ball, straws, and clay or putty so they can make a globe with an axis (one straw sticking out of the "north end" of the ball and the other one out of the "south end"). Tell them they can use this to help them visualize what is happening

with each planet. One member of the group can stand in the center of the group and be the Sun and another member can orbit the ball around the Sun to show what is happening during the course of that planet's year.

6. Now give each group a copy of the "Characteristics of the Planets" sheet (showing the axis tilt, eccentricity, and length of the year for each planet). Assign one or more planets to each group. Ask them to discuss and then write down whether each planet has seasons and why. (If the class has studied the planets and they know which ones have an atmosphere and which don't, you can tell them to ignore the effect of any atmosphere for purposes of this discussion and just focus on axis tilt and eccentricity.)

7. After the groups have each reached a consensus, have a class-wide symposium about the seasons on a few or all of the other planets. Have one group lead the discussion for each planet. Provide "correct" answers (see the Teacher Version of the "Characteristics of the Planets") as late as possible in the discussion.

   Teacher note: Some of the seasonal effects, especially the ones summarized in the notes on Mars and Uranus, may be too complicated for younger students. It may be best to stick to the simpler situations, such as those on Venus, Jupiter, and Saturn.

8. To check how well students understand these results, you can (for homework, if there isn't time in class) ask each student to pick one planet and tell you if it has seasons, and if so, how long each season is.

Characteristics of the Planets

TEACHER VERSION*

Planet	Are there seasons? Why or why not? Explain.
Mercury	Mercury has no axis tilt, but the eccentric orbit leads to complex differences in temperature that give the planet its seasons;. This is complicated by the fact that a Mercury day is 58.67 Earth days long, exactly ⅔ of the length of Mercury's year.
Venus	Neither factor is significant, and there are no seasons.
Earth	The axis tilt is the dominant factor; eccentricity is small.
Mars	The axis tilt is the dominant factor, but eccentricity is also significant. This means the two hemispheres (north and south) on Mars have somewhat different seasons (see note below).
Jupiter	Neither factor is significant, so there are no real seasons.
Saturn	The axis tilt is the dominant factor. There are definite seasonal changes over the course of a Saturn year, which takes 30 Earth years.
Uranus	The axis tilt is the dominant factor. A tilt of 90° would mean Uranus is sideways—the tilt is so close to 90°s that the planet is essentially spinning on its side, making seasonal differences complicated (see Notes on pp. 130–131).
Neptune	The axis tilt is dominant factor in causing seasons.

*A copy of the student handout is available at www.nsta.org/solarscience.

Characteristics of the Planets

Planet	Axis tilt	Eccentricity (goes from 0 to 1)	Length of year (in Earth years)	Are there seasons? Why or why not? Explain.
Mercury	0°	0.21	0.24	
Venus	3°	0.01	0.62	
Earth	23.5°	0.02	1	
Mars	24°	0.09	1.88	
Jupiter	3°	0.05	11.9	
Saturn	27°	0.06	29.7	
Uranus	98°	0.05	84.3	
Neptune	29°	0.01	165	

Notes

MARS

The seasons are the most complicated on Mars because *both* the tilt of 24° and the changing distance from the Sun (larger eccentricity) play a role. Mars is farthest from the Sun when it is winter in the Southern Hemisphere and summer in the north. It's closest to the Sun when the south has summer and the north has winter. So, seasonal differences in the south are more extreme. (Put another way, the two effects reinforce each other in the south but work against each other in the north.)

URANUS

The planet orbits on its side. That is, the axis around which Uranus rotates is roughly in the same plane as the planet's orbit. This leads to strange seasons. At the height of summer for the Northern Hemisphere, for example, the axis points almost directly toward the Sun. This means the northern side of the planet is in continuous sunlight while

FIGURE 2.12

Axis tilts on the planets in our solar system

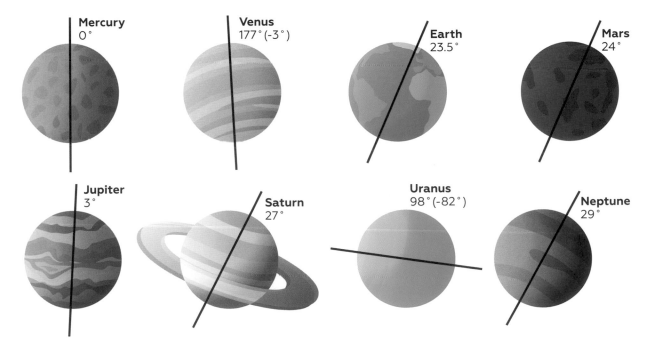

Understanding and Tracking the Annual Motion of the Sun and the Seasons

the southern side is in continuous darkness. When it is summer in the Southern Hemisphere, the situation is reversed. In between, during spring and fall, the planet is sideways to the Sun, so both hemispheres get night and day (sunlight and darkness) every 17 hours. When the Voyager 2 spacecraft flew by Uranus in 1986, the South Pole of the planet was pointed at the Sun (see Figures 2.12 and 2.13 and Table 2.4).

FIGURE 2.13

Uranus's seasons

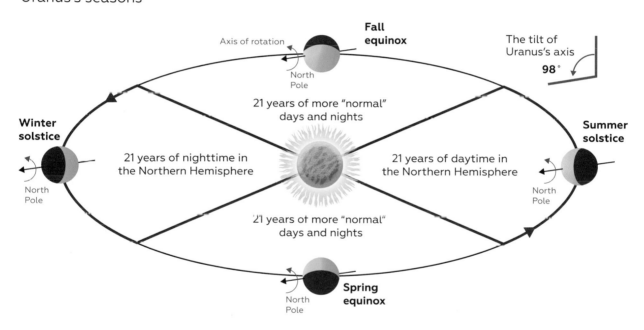

TABLE 2.4

Years of Uranus's seasons

Northern Hemisphere	Year	Southern Hemisphere
Winter solstice	1986	Summer solstice
Spring equinox	2007	Fall equinox
Summer solstice	2028	Winter solstice
Fall equinox	2049	Spring equinox

I Can't Make It Come Out Even

FITTING DAYS AND YEARS INTO A WORKABLE CALENDAR

Overall Concept

Students are told the length of an Earth year, an Earth month, and an Earth day and are challenged to make a consistent calendar. After learning about the different calendar systems we have devised on Earth, they are given data for an imaginary planet, Ptschunk, and have to apply their knowledge to create a calendar for that world.

Objectives

Students will

1. understand the origin of various units of time that are based on the astronomical cycles visible in the Earth's sky,

2. devise their own calendar for Earth that uses these cycles, and

3. apply what they learned to making a calendar for a fictional planet with different cycles.

Advance Preparation

1. The cycle of the Moon is covered in Chapter 4 of this book, and you may want to look ahead to the Content Background of that chapter (pp. 276–287). Students who have not yet studied the phases of the Moon may need some background information from you about the 29.5-day cycle of the Moon's phases.

2. If you teach younger students, you may want to modify the "Planet Ptschunk Worksheet" to list fewer characteristics for them to work with.

MATERIALS

One per group:

*   "Calendar Worksheet" (p. 137)

*   "Planet Ptschunk Worksheet" (p. 138)

One per student:

*   Calculator

*   Astronomy lab notebook

Procedure

1. Begin a discussion with students about how we keep track of time today. Their answers should include watches, phones, calendars, computer screens, and so on. Ask students to think about what it was like for people thousands of years ago, before we had clocks, watches, or electricity. They still needed to keep track of time to figure out when to serve meals, to be ready for the changes of the seasons, to plant and harvest crops at the right time, and so on. The astronomical cycles we have been discussing in this chapter and the last were natural markers for keeping track of time.

2. Ask students what astronomical cycles on Earth repeat in a noticeable way and can help us keep track of time. They should suggest days, months, and years in the discussion.

   Teacher note: If students bring up a week, ask them to justify this suggestion by saying what astronomical cycle they have in mind. Use this question to let them conclude that there is nothing in the sky that happens in a week. Ancient people made up the time unit of a week of seven days to honor the seven "gods" they saw moving in the sky among the stars (the Sun, the Moon, Mercury, Venus, Mars, Jupiter, and Saturn). That's why our weekend has a "Saturnday" and a Sunday, and school starts on "Moonday." (In English, the other four days get their names from Viking gods instead of Roman ones. The other day names are more like the planets' names in Spanish or French; so Wednesday is miercoles in Spanish and mercredi in French, after the planet Mercury.)

3. Work with the students to get numbers on the board and into their notebooks for each cycle:

   • One day is how long it takes for the Earth to turn on its axis, so the Sun appears to return to the same position in the sky (1 day = 24 hours).

   • One month is how long it takes the Moon to go through a full cycle of its phases, for example from full Moon to the next full Moon, about 29.5 days.

- One year is how long it takes for the Earth to go through the full cycle of its seasons (or to orbit the Sun once; 1 year = 365.25 days).

   Teacher note: Because the Earth is orbiting the Sun while it spins, the time for the Sun to return to the same position in the sky is four minutes longer than the time for the Earth to spin once relative to the stars, but at the middle-school level we can ignore that difference in our discussions.

4. Once the students have these lengths of time in their notebooks, ask them which of these cycles would ancient people have been most aware of (which is easiest to notice)? *It is the day, with the day–night cycle being hard to miss.*

5. Ask the students if a day goes evenly into one lunar cycle or into the year. Does the month go evenly into the year? (Encourage them to use their calculators.) *The answer to both questions is no.* That is the challenge of the calendar—how to make astronomical cycles that are not even multiples of each other fit into a workable system. Tell the students that human beings have been struggling with this issue for thousands of years.

6. Have the students work in groups of three or four. Give each group the calendar handout and have them discuss how to make a calendar that has days and months during the year and still comes out even. Encourage them to use their calculators as needed. Some students may know the answer about the calendar we use already, but tell them they can also try other ways of doing it besides the one we have adopted. (For example, there could be some days at the end of the year that are not part of any month but are a vacation for everyone.) The main point is for students to grapple with the issue, not to get any particular "right" answer.

7. Have the groups share their ideas with the whole class. Then finish by discussing our modern calendar system, in which

    - there are 12 months of varying length (28 to 31 days) in the year, adding up to 365 days in 12 months;

    - the cycle of the Moon's phases in our sky is not lined up with the calendar months, and any specific lunar phase

(e.g., full Moon) can be on any day of a given calendar month; and

- to take care of the quarter day in the 365.25 day length of the year, we make every fourth year a *leap year*, with 366 days in it (adding a February 29 in that year).

Teacher note: The length of the year is not exactly 365.25 days. To compensate for this, century years are treated differently. Only century years divisible by 400 are leap years, the others are not. So 1900 was not a leap year, 2000 was, and 2100, 2200, and 2300 will not be. This minor correction can be omitted at the middle school level.

8. Share with the students one other solution to devising a calendar, the lunar calendar. See the part about lunar calendars in the Content Background of this chapter (p. 77).

Teacher note: In some lunar-cycle based calendars, a new month begins when the smallest crescent Moon that your eyes can see appears in the sky; in others it begins at the time of the new Moon. Since 29.5 days doesn't divide evenly into 365.25 days, the months of a purely lunar calendar get out of step with the cycle of the seasons that makes the year. To adjust for this, some cultures added an extra month from time to time to keep the lunar system aligned with the time when the seasons occur. Such combined calendar systems are called *lunisolar*.

9. Now tell students that it's the future, and astronomers have discovered a planet around another star, the time cycles of which we have been able to measure. We'll call it Ptschunk (pronounced "puh-choonck"). Its characteristics are different from Earth's:

- Ptschunk's spin rate is 21.3 Earth hours.

- The planet's revolution about its sun takes exactly 302 Ptschunkian days.

- Ptschunk has two bright moons of about equal size, one taking 20 Ptschunkian days to go through its cycle of phases, the other 60 days.

- Three other planets are visible in Ptschunk's sky (in addition to the sun and the two moons).

10. Give students the "Planet Ptschunk Worksheet" and ask them (in their groups) to make a calendar system for Ptschunk and be prepared to explain it. Encourage them to come up with some important Ptschunkian holidays (they can be creative). There is no right answer here; students should be rewarded for good thinking and can approach this part of the experience with a sense of humor. (Remember, you can modify that worksheet before distributing it to include fewer characteristics for students to worry about.)

11. (*Optional*) If you have time, have students consider how to make a calendar for Mars. A day on Mars (called a sol) is equal to 1.026 Earth days. A year on Mars is 686.98 Earth days (1.88 Earth years). It follows that Mars rotates 669.6 times on its axis during each orbit around the Sun.

    Mars has two tiny moons. Phobos takes 7 hours and 39 minutes to orbit Mars, while Deimos takes 30 hours and 18 minutes. Both are so small, you would need a telescope to see phases. If you could see Deimos' phases, for example, they would repeat on a cycle that takes 1.2648 Earth days. So, the moons are probably not much good for defining months that are significantly different from days.

    Since Mars takes about twice as long to orbit the Sun as the Earth does, seasons there are around double the length of Earth's. From this information, can students suggest a calendar for Mars? (Other people have tried this: See, for example, the following:

    • *http://pweb.jps.net/~tgangale/mars/mst/darian.htm*

    • *http://pweb.jps.net/~gangale3/other/millenn.htm*

    • *http://planetary.org/explore/space-topics/mars/mars-calendar.html, and http://digilander.libero.it/vcoletti/ideas/marscalendar.html.*)

12. (*Optional*) Have students research the calendar used by another culture. For example, they can look at ancient Egyptian or Mayan calendars. Or they can look at the older Hebrew, Islamic, Chinese, or Japanese calendars, which used the lunar cycle as the key time indicator. Students can do this individually or in small groups, and one representative of each calendar system can then report to the entire class on what their research revealed. See the list of resources about the calendars of other cultures in the Resources for Teachers section of this chapter.

Calendar Worksheet

※

Your name: _____ Date: _____

Others in your group: _____

Our system of keeping track of units of time is based on astronomy. Ancient people looked at the sky and noticed *cycles* (events that repeated on a regular schedule).

- One day is how long it takes for the Earth to turn on its axis, so the Sun appears to return to the same position in the sky (1 day = 24 hours).

- One month is how long it takes the Moon to go through a full cycle of its phases, for example from full moon to the next full moon, about 29.5 days.

- One year is how long it takes for the Earth to go through the full cycle of its seasons (or to orbit the Sun once; 1 year = 365.25 days).

- One week is a made-up unit of seven days (based on the seven wandering objects in the sky visible to the ancients: the Sun, the Moon, and five visible planets).

Try dividing the length of the lunar cycle into the year. Does it come out even? Do any of the cycles divide evenly into any of the other cycles?

How can we make a yearly calendar that includes at least days and months and works so that it can start over again after 365.25 days? Discuss any and all ideas group members have and write your suggestions below:

_____

_____

_____

_____

_____

_____

_____

_____

Planet Ptschunk Worksheet

Astronomers have discovered a new planet orbiting another star, called Ptschunk (pronounced "puh-choonck"). Its characteristics are different from Earth's:

- Ptschunk's spin rate (its day) is 21.3 Earth hours.

- The planet's revolution about its sun (its year) takes exactly 302 Ptschunkian days.

- Ptschunk has two bright moons of about equal size, one taking 20 Ptschunkian days to go through its cycle of phases, the other 60 days.

- Three other planets are visible in Ptschunk's sky (in addition to the sun and the two moons).

Remember that on Earth we define the cycles of time we use as follows:

- One day is how long it takes for the Earth to turn on its axis, so the Sun appears to return to the same position in the sky (1 day = 24 hours).

- One month is how long it takes the Moon to go through a full cycle of its phases, for example from full moon to the next full moon, about 29.5 days.

- One year is how long it takes for the Earth to go through the full cycle of its seasons (or to orbit the Sun once; 1 year = 365.25 days).

- One week is a made-up unit of seven days (based on the seven wandering objects in the sky visible to the ancients: the Sun, the Moon, and five visible planets.) Nothing happens in the sky in one week.

Please make a calendar system for Ptschunk and be prepared to explain it. If you have time, come up with some important Ptschunkian holidays (be creative).

_____

_____

_____

_____

_____

EXPERIENCE 2.9

Write a Picture Book for Kids

Overall Concept

Each student writes and illustrates a picture book for a younger student that explains how the relationship between the Earth and Sun causes the seasons. Three writing prompts are provided to encourage the inclusion of key ideas learned in the chapter. If students already wrote a picture book as part of Experience 2.5, "Reasons for the Seasons Symposium" (see p. 97), then you may want to have them revise their books; there is often significant revision required. You may find it better to ask the students to start fresh and write a second edition of their book that incorporates the new learning from the research they did as part of this chapter.

Objective

Students will be able to clearly explain the various factors that cause the seasons.

Procedure

1. Tell students that they will write and illustrate a children's picture book for students two to three years younger than they are. The book must include

    a. explanations and illustrations of how the distance to the Sun, the tilt of the Earth's axis of rotation, and the Earth's orbiting the Sun affect our seasons;

    b. how the motion of the tilted Earth around the Sun affects the amount of daylight we receive and how it affects the height of the Sun above the horizon; and

MATERIALS

*   Blank pages for a book (e.g., pages in the student's astronomy lab notebook, sheets of paper that get stapled in the corner, or a book made from 8 ½ × 11 sheets of paper folded in half)

*   Colored pencils or pens (enough so all students can have a selection), plus access to magazines, images on the web, or other resources for use in illustrating their books

c. how this applies to a person in the Southern Hemisphere, who is at the same distance south of the equator as you are north of the equator.

2. Key elements to look for in the finished books include the following:

a. The 23.5° tilt of the Earth's axis relative to the direction of the Sun as the Earth orbits the Sun means that sometimes the Northern Hemisphere is tilted "toward" the Sun (summer solstice, around June 21) and sometimes tilted "away" from the Sun (winter solstice, around December 21). At the equinoxes, neither hemisphere is tilted toward or away from the Sun.

b. Around the summer solstice, the Sun appears higher in the sky and the sunlight is more concentrated for those people in the Northern Hemisphere. The days also last longer, which increases the amount of heating that occurs each day. Around the winter solstice in the Northern Hemisphere, the days are shorter and the Sun's light is most spread out, so there is less heating. At the equinoxes, all parts of the Earth experience 12 hours of daylight and 12 hours of night, and everything is more equal.

c. When it is summer in the Northern Hemisphere, it is winter in the Southern Hemisphere. When it is winter in the Northern Hemisphere, it is summer in the Southern Hemisphere.

If there is a need to give a score for the content of the book, consider one to two points for each of the three elements listed above, depending on the thoroughness of the response.

EXPERIENCE 2.10

E-mail Response to "How Can This Be True?"

Overall Concept

Each student prepares an e-mail response to the teacher's friend in *engage* Experience 2.2, "How Can This Be True?" explaining why the Southern Hemisphere can experience summer while the Northern Hemisphere experiences winter.

Objective

Students will be able to clearly explain what they have learned about the reasons for the seasons.

Procedure

1. Tell students that you would like them to write (in their astronomy lab notebook) a response to your friend who sent you the two e-mails in the "How Can This Be True?" experience from the beginning of the chapter. Unlike Experience 2.9, "Write a Picture Book for Kids," this experience provides limited prompts to encourage student's ability to decide what elements are important to include in the message. The only prompt you should provide is that the message must include information learned during the study of the Earth's relationship with the Sun.

2. Key elements to look for in the finished books include the following:

    • The tilt of the Earth's axis relative to the direction of the Sun as the Earth orbits the Sun means that sometimes the Northern Hemisphere is tilted toward the Sun (summer solstice, around

MATERIALS

One per student:

• Astronomy lab notebook

• (*Optional*) Computer with word processing or e-mail program if you wish to have students respond electronically

June 21), and sometimes tilted away from the Sun (winter solstice around, December 21).

- Around the winter solstice, this tilt of the Earth's axis causes the Sun to be lower in the sky in the Northern Hemisphere compared with locations in the Southern Hemisphere. This means the sunlight is less concentrated for people in the Northern Hemisphere. In addition, the days are shorter than for people in the Southern Hemisphere.

- When it is summer in the Northern Hemisphere, it is winter in the Southern hemisphere. When it is winter in the Northern Hemisphere, it is summer in the Southern Hemisphere. This explains why in December it can be beach weather in Sydney in the Southern Hemisphere while it is winter weather in New York in the Northern Hemisphere.

EXPERIENCE 2.11

Reasons for the Seasons Revisited

Overall Concept

Students label a diagram that shows the relationship of the Sun and Earth at different times of year and describe how this explains the reason for the seasons.

Objective

Students will be able to explain clearly what they have learned about the reasons for the seasons.

Procedure

These same instructions are on the worksheet for students to reference as they complete the experience.

1. Have students cut off the top half of the "Reasons for the Seasons Revisited Worksheet" and attach it to a blank page in their astronomy lab notebooks.

2. Have them cut out the four images of the Earth at the bottom of the worksheet and attach each one at an appropriate place on the top half of the worksheet. Remind them to show the correct tilt of the Earth's axis for each location.

3. Have them label the appropriate season for each location where they placed their Earths. Have them make sure they note which way the Earth is orbiting the Sun.

4. At the bottom of the page, have the students explain how the diagram shows the primary reasons for the seasons on the Earth.

MATERIALS

One per student:

- "Reasons for the Seasons Revisited Worksheet" (p. 144)

- Tape or glue

- Astronomy lab notebook

Reasons for the Seasons Revisited
Worksheet

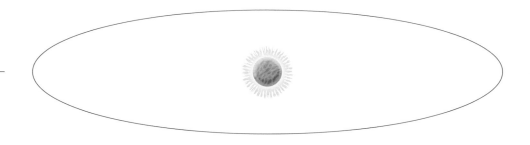

1. Cut off the top half of the "Reasons for the Seasons Worksheet" and attach it to a blank page in your astronomy lab notebook.

2. Cut out the four images of the Earth and attach each one at an appropriate place on the Earth's orbit on the worksheet in your notebook. Be sure the Earth's axis is tilted the proper amount and direction for each location. Darken the nighttime side of the Earth for each location.

3. Label the appropriate season in the Northern Hemisphere for each location where you placed your Earths. Make sure you note which way the Earth is orbiting the Sun.

4. At the bottom of the page in your notebook, explain how the diagram shows the reasons for the seasons on the Earth.

Reasons for the Seasons Revisited
Worksheet Answers

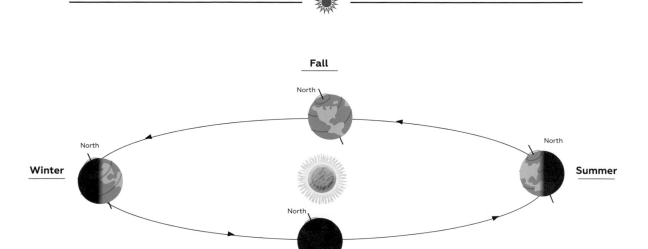

Summer and winter are labeled correctly for the positions on the right or left. Spring and fall are labeled properly in the front and back. The tilt of the Earth's axis is in the proper orientation and an arrow shows which way the Earth is orbiting the Sun. The explanation at the bottom of the page should include some or all of the items identified in the answer to evaluate Experience 2.9, "Write a Picture Book for Kids," especially the first item about the tilt of the Earth's axis relative to the Sun. The side of the Earth away from the Sun is darkened.

EXPERIENCE 2.12

What Do We Think We Know? Revisited

Overall Concept

Students revisit their responses to *engage* Experience 2.1, "What Do We Think We Know?" allowing them to see how their understanding of the key concepts in the chapter has changed.

Objective

Students will be able to explain clearly what they have learned about the reasons for the seasons.

Procedure

1. Have the students label the top of a page in their astronomy lab notebook, "What I Now Know." Right below that, have them write the first and second question you want them to consider and discuss: "Does the Sun's position in the sky at noon change throughout the year? If so, how does it change, and what causes the change?" On the following page, have them write the next question at the top: "What causes the differences we experience with the changing seasons, such as length of daylight and temperature?"

2. If you are not using this information to assess individual student understanding, then use the think-pair-share process described in the Introduction to this book (p. xix) to (1) have each student individually write how his or her thinking has changed; (2) discuss their comments with other students in small groups and add more detail to their notebooks as desired; and finally, (3) update, during a whole-class discussion, the answers collected at the beginning of the unit.

MATERIALS

For the class:

*   Whiteboard, blackboard, or poster paper where student preconceptions were recorded at the beginning of the unit

One per student:

*   Astronomy lab notebook

3. If you want to use this information to assess each student, then you will want to collect the notebooks for review before the discussion in small groups. After you review the notebooks, it is still important to have the small-group and whole-class discussions that summarize what everyone learned.

Video Connections

- *Bill Nye the Science Guy Explains the Seasons.* The first third of this half-hour program is devoted to a comedic explanation of the seasons, with several experimental demonstrations that help clarify the material in this chapter: *www.youtube.com/watch?v=w1qTQXN84Ko*

- *Seasons, Sun Angle, and Latitude.* A lecture with animations on the various effects that cause the seasons and how they vary with latitude (10 min.): *www.youtube.com/watch?v=lnZ7ldOlxxU*

- *Teach Astronomy: Early Roman Calendar.* A video with astronomer Chris Impey of the University of Arizona, explaining some of the history of calendars (2 min.): *www.youtube.com/watch?v=etrbBXo2Fyk*

- Follow this with some of the other short videos from Chris Impey's Teach Astronomy project:

  - Julian Calendar: *www.youtube.com/watch?v=358fY7vb-44*

  - Modern Calendar (the Gregorian reform): *www.youtube.com/watch?v=KOd4vUj2fGM*

  - Solar and Lunar Calendars: *www.youtube.com/watch?v=plkAvc1vhGc*

Math Connections

- To develop graphing skills, students could plot sunrise and sunset for their home location (or for any other location on Earth) and actually make a graph that shows one of the key aspects of the seasons. Students can find such information for a year here:

  - *http://aa.usno.navy.mil/data/docs/RS_OneYear.php*

  - *www.timeanddate.com/worldclock/sunrise.html* (first put in your location, then, when it gives you the information for the current month, you can get other months from the results page)

  - *www.esrl.noaa.gov/gmd/grad/solcalc*

- Some math connections related to the calendar can be found already in Experience 2.8, particularly in the optional parts (such as making a calendar for the planet Mars.)

- Students can consider the following scenario. What if the Earth took not 365.25 days to go around the Sun but 300 days exactly? Assume the month and the day remain the same length they are now. How would the seasons change? (*Hint:* For example, how long a summer vacation might you have?) What kind of calendar might we adopt?

- A short discussion of how to calculate what year is a leap year and what year isn't can be found on the following Windows to the Universe page: *www.windows2universe.org/earth/leap_year.html*

- Another activity about calendars, with more advanced math skills required, is PUMAS (JPL Activities), "How Many Days in a Year?": http://pumas.jpl.nasa.gov/examples/index.php?id=46

- Formulas for figuring out what day of the week any date falls on and what future years you can reuse this year's calendar can be found here: www.tondering.dk/claus/cal/chrweek.php

Literacy Connections

- Harper, D. A brief history of the calendar. www.obliquity.com/calendar.

- Moving to Alaska. Weather in Alaska. www.alaska.net/weather.html.

- American Museum of Natural History, Rice University, and the Education Development Company, Inc. Letter from Stephanie: Day and night cycles in Antarctica. www.amnh.org/education/resources/rfl/web/antarctica/s_cycles.html.

- EarthSky. What are the seasons like on Uranus? http://earthsky.org/space/what-are-the-seasons-like-on-uranus.

- SciJinks. What is the solstice? NASA. http://scijinks.jpl.nasa.gov/solstice.

Cross-Curricular Connections

1. The seasons play a big role in life on Earth and in stories we tell, movies we make, and music we enjoy. Ask students (in small groups or as a homework assignment) to come up with examples in which the seasons and their differences are important in a story, a movie, or a musical piece.

(*Hint:* Students who know classical music may be able to find several pieces called "The Seasons" by Vivaldi, Tchaikovsky, Glazunov, and Haydn.)

2. If your students are studying poetry in their English class, perhaps the English teacher can help them find examples of poems in which the weather in different seasons plays a role. Each group might bring in an example of poetry about one of the four seasons. Lists of seasonal poems have been prepared on the web if you want to help students find poems easily. See, for example, the series created by Bob Holman and Margery Snyder:

- Autumn poems: http://poetry.about.com/od/ourpoemcollections/a/autumnpoems.htm

- Winter poems: http://poetry.about.com/od/ourpoemcollections/a/winterpoems.htm

- Spring poems: http://poetry.about.com/od/ourpoemcollections/a/springpoems.htm

- Summer poems: http://poetry.about.com/od/ourpoemcollections/a/summerpoems.htm

3. You might ask students to read an article about the reform of the old calendar when our modern system (called the Gregorian calendar) was introduced, replacing the old system that had been in place since the Roman emperors (called the Julian calendar). By the time England got around to making the change, eleven days had to be removed

from one year to get the seasons back where they belonged. See, for example, *http://mentalfloss.com/article/51370/why-our-calendars-skipped-11-days-1752*.

4. For the most comprehensive website for understanding all the cross-curricular aspects of the calendar, including women's calendars, historical calendars, calendars and technology, and much more, we recommend Calendar Zone (*www.calendarzone.com*).

5. Keeping track of the seasons was an activity of vital importance in ancient cultures. The people needed to know when to plant, when to reap, and when it was time to prepare for more extreme weather. Many archaeological sites around the world are oriented to be "calendars in stone"— among them Stonehenge in England and the Temple of Kukulkan in Chichén Itzá in Mexico. Students can research these and other astronomically oriented sites and report on how they helped their builders keep track of the seasons.

   For some useful resources, see the books by Anthony Aveni in the Resources for Teachers section at the end of the chapter as well as the following:

- A fine guide to sites around the world, written for beginners with humor and verve:
    Krupp, E. C. 1997. *Skywatchers, shamans, and kings: Astronomy and the archaeology of power.* New York: John Wiley and Sons.

- A good, nontechnical introduction to the myths, constellation, calendars, astronomical buildings, and world views of various cultures:
    Penprase, B. E. 2011. *The power of the stars: How celestial observations have shaped civilization.* New York: Springer.

- An introduction to ancient sites where the movements of celestial objects were tracked over the years (with a special focus on tracking the Sun):
    Stanford Solar Center. Ancient observatories, timeless knowledge. *http://solar-center.stanford.edu/AO*.

- This site, produced by the education group at the Berkeley Space Sciences Lab, connects Maya knowledge to modern astronomy:
    University of California, Berkeley, Center for Science Education at the Space Sciences Laboratory. Calendar in the sky. *www.calendarinthesky.org*.

- The NASA Sun-Earth Connection Education Forum site offers virtual visits to Mayan astronomical sites and Chaco Canyon placed in appropriate historical, cultural, and scientific contexts:
    NASA. Traditions of the Sun. *www.traditionsofthesun.org*.

- An art historian examines Stonehenge from many perspectives, including the astronomical.
    Witcombe, C. Archaeoastronomy at Stonehenge. *http://arthistoryresources.net/stonehenge/stonehenge.html*.

Resources for Teachers

THE SEASONS

- A book of activities and information from the Great Explorations in Math and Science curriculum project at the Lawrence Hall of Science and NASA:
  Gould, A., C. Willard, and S. Pompea. 2000. *The real reason for the seasons: Sun–Earth Connections*. Berkeley, California: GEMS.

- A review of research about learning and teaching the seasons:
  Sneider, C., V. Bar, and C. Kavanagh. 2011. Learning about the seasons: A guide for teachers and curriculum developers. *Astronomy Education Review* 10 (1). *http://portico.org/stable?au=pgg3ztf87gm*

- A quick overview of the seasons by amateur astronomer Ed Hitchcock:
  The Budget Astronomer. The tilt of the Earth and the seasons. *www.budgetastronomer.ca/index.php?page=th-tilt-of-the-earth-and-the-seasons.*

- An activity to make an Earth-centric model that shows the Sun's apparent motion across the sky for different seasons:
  Scherrer, P., and D. Scherrer. Solstice and equinox ("suntrack") season model. Stanford Solar Center and NASA. *http://solar-center.stanford.edu/activities/Suntrack-Model/Suntrack-Model.pdf.*

- An interactive animation of the seasons:

Fix, J. D. 2006. Seasons interactive. McGraw-Hill Higher Education. *http://highered.mheducation.com/sites/007299181x/student_view0/chapter2/seasons_interactive.html*

- Two websites that explain seasons on other planets in some detail:

  - NASA. Weather, weather, everywhere? *www.nasa.gov/audience/foreducators/postsecondary/features/F_Planet_Seasons.html.*

  - NASA. Interplanetary seasons. *http://science.nasa.gov/science-news/science-at-nasa/2000/interplanetaryseasons.*

- Information about Uranus and its seasons:
  Goldstone Apple Valley Radio Telescope Project. Understanding Uranus. *www.lewiscenter.org/documents/Global%20Programs/Uranus_Intro.pdf.*

UNITS OF TIME AND THE CALENDAR

BOOKS AND ARTICLES

Aveni, A. 2002. *Empires of time: Calendars, clocks, and cultures*. Boulder, CO: University Press of Colorado.

Aveni, A. 1997. *Stairways to the stars: Skywatching in three great ancient cultures*. New York: John Wiley and Sons.

Barnett, J. E. 1999. *Time's pendulum: From sundials to atomic clocks, the fascinating history of timekeeping and how our discoveries changed the world*. New York: Harcourt Brace.

Bartky, I., and Harrison, E. 1979. Standard and daylight-saving time. *Scientific American* 240 (5): 46–53.

Duncan, D. E. 2001. *Calendar: Humanity's epic struggle to determine a true and accurate year.* New York: Avon Books.

Steel, D. 2000. *Marking time: The epic quest to invent the perfect calendar.* New York: John Wiley and Sons.

WEBSITES

WebExhibits. Calendars through the ages. *www. webexhibits.org/calendars/index.html.* (see under the menu bar "Various Calendars" to get good explanations of the calendars of other cultures, including Chinese, Islamic, and Jewish.

Tøndering, C., and T. Tøndering. The calendar FAQ. *www.tondering.dk/claus/calendar.html.*

NASA Earth Observatory image of
Asia and Australia at night

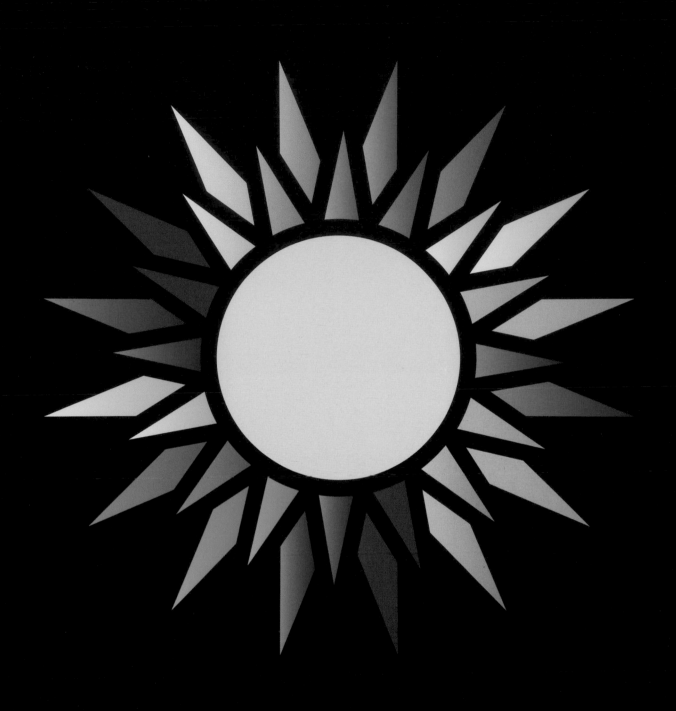

3

Solar Activity and Space Weather

The Sun's light is the ultimate source of energy for human civilization (Figure 3.1). But our home star also puts out a steady wind and occasional storms of charged particles, both of which can interact with natural and human processes on Earth. As the Sun rotates, we see spots (darker areas on the Sun's surface) move across the face of our star, which allows us to measure its rate of rotation. Astronomers have learned from counting these sunspots that the Sun undergoes an 11-year cycle of rising and falling activity. When the number of spots is high, the Sun is more likely to have *flares*—outpourings of radiation and high-speed particles. Sometimes, there are great storms of charged particles that leave the Sun in a burst; these are called *coronal mass ejections* (CMEs).

The more powerful solar storms can disrupt power grids on Earth, damage delicate electronics in orbiting instruments, and endanger astronauts on the International Space Station (ISS). As our civilization increasingly depends on connections among regional power networks and on interactions with satellites in space, our students' generation will need to have a good understanding of the activity on the Sun, the "space weather" it causes, and how these can affect our lives.

FIGURE 3.1

The brilliant Sun, a crescent Earth, and the International Space Station are seen in this 2009 photo by an astronaut.

Learning Goals of the Chapter

After doing these activities, students will understand the following:

1. The Sun, like the Earth and the Moon, rotates, taking roughly 27 days to spin once. (Because the Sun is not solid, its layers at different latitudes take different amounts of time for a full turn.)

2. The energy emerging through the Sun's surface, together with the Sun's magnetism and rotation, produce violent, churning solar activity in its outer layers. These layers include the visible surface, called the *photosphere*, and the Sun's atmosphere, which contains the *chromosphere* and the *corona*. The Sun's activity is seen as *sunspots*, *prominences*, *flares*, and CMEs.

3. Solar activity varies on a roughly 11-year cycle.

4. Space weather near the Sun and near the Earth is the result of both a steady flow of charged particles from the Sun moving away in all directions (the solar wind) and sudden bursts of energetic particles (flares and CMEs). Particles arriving from the Sun interact with the Earth and its magnetic field (the *magnetosphere*).

5. Just like storms on Earth can be dangerous, strong "gusts" of space weather can be harmful to sensitive technology in space and on Earth, making it useful to have advance space weather predictions and ongoing space weather reports.

6. In addition to visible light and atomic particles (mostly electrons and protons), the Sun also puts out invisible radiation, especially *infrared* and *ultraviolet* (UV) rays.

Overview of Student Experiences

Teacher note: Not all the experiences in this chapter need to be done during a unit about the Sun, and these experiences can certainly be done in a different order than the one we present. We simply want to make a wide range of experiences in this fascinating area available for our readers to choose among.

ENGAGE EXPERIENCES

- **3.1. What Do We Think We Know?** Students consider and discuss their thoughts, first in small groups and then with the whole class regarding the following questions: (1) What kind of features can we observe on the surface and in the atmosphere of the Sun? (2) Does the Sun rotate, and if so, how can we tell?

- **3.2. Be a Solar Astronomer:** Students look at a number of images of the Sun taken a few days apart. They make lists of features they can see, what changes have occurred over the days, and any questions they have.

EXPLORE EXPERIENCES

- **3.3. Safe Solar Viewing: Project and Record Your Own Images of the Sun:** After receiving safety instructions, the class uses binoculars to project an image of the Sun and begins making daily drawings of its appearance. Internet images can be substituted for cloudy days, days when no sunspots are visible, or if an instrument for projecting a magnified view of the Sun is not available. (Other methods of observing the Sun safely are given at the end of the activity.)

- **3.4. Discover the Sunspot Cycle:** Students work in small groups to count sunspots on representative images of the Sun. Students then plot data over several decades to see the pattern of increasing and decreasing numbers, leading them to discover the 11-year sunspot cycle.

EXPLAIN EXPERIENCES

- **3.5. How Fast Does the Sun Rotate?** Students think about how we could tell that a ball of hot gas like the Sun rotates. Then they use sequential images of the Sun's disk with sunspots to measure the rotation period of the Sun.

- **3.6. Space Weather: Storms From the Sun:** Students brainstorm and discuss what goes into weather reports on Earth (e.g., temperature, wind, gusting rain or snow, etc.) and how the weather reports (online, on TV, in newspapers) are put together. After learning more about the Sun, they make a list of space weather phenomena on the Sun and how they might affect the Earth. Either in small groups in a classroom equipped with computers and online access or individually at home, students search for websites with accessible reports on space weather. (If needed, a list of good websites is also provided at the end of the activity.) In a class discussion, groups compare the results of their research. Students watch a NASA video on a CME that missed the Earth and read about the Carrington event and other times when a CME did hit the Earth. Then, in small

groups, they brainstorm about the effects that a solar storm could have on the Earth. The class comes together in a symposium to discuss ideas.

- **3.7. What Else Cycles Like the Sun?** Students work in small groups looking at two different types of data from the Sun and Earth (geomagnetic activity and the S&P 500 stock index) to see whether each phenomenon correlates with sunspot cycles. Students then discuss which phenomena might be linked to the 11-year solar activity cycle and which are not.

ELABORATE EXPERIENCES

- **3.8. The Multicolored Sun: An Introduction to Electromagnetic Radiation:** Students begin with the challenge of understanding how their TV remote communicates with their TV. They turn off the light and discover that they can't see the beam that reaches the TV. From this, they are introduced to the idea that there is a spectrum of "invisible light" or electromagnetic radiation. They discuss any previous knowledge they have of these other bands of the spectrum. Then, in small groups, they examine and discuss images of the Sun taken the same day but in several different bands of the electromagnetic spectrum (e.g., x-rays, UV, and visible light). They discuss why the pictures look so different and then come together in class discussion with further guidance from the teacher.

- **3.9. Student Detectives and the Ultraviolet Sun:** Students continue their exploration of the spectrum by examining

the UV radiation coming from the Sun using UV-sensitive beads. Student groups are given the challenge of seeing how much different substances block UV waves, and the groups report back to the whole class on what they find.

- **3.10. Additional Ways of Observing the Sun Safely:** Students learn and practice other methods for safe Sun viewing. These are particularly useful for times when a partial or total solar eclipse is visible.

EVALUATE EXPERIENCES

- **3.11. Space Weather Report:** Students in small groups pretend they are auditioning for the evening TV news and produce a brief weather report segment about current space weather in the style of those newscasts. Students are provided with (or find) websites for getting current information, but it's up to them to write a report or make a presentation that ordinary people watching TV news can understand.

- **3.12. Predict the Next Sunspot Maximum and Minimum:** Using the information from *explore* Experience 3.4, students predict when the next maximum time and the

next minimum time for sunspots will be. Students can, if you want, also predict what other phenomena will be at maximum when the Sun is most active (from *explain* Experience 3.6).

Recommended Teaching Time for Each Experience

Table 3.1 shows the amount of time needed for each experience.

Connecting With Standards

Table 3.2 (pp. 160–161) shows the *Next Generation Science Standards* (*NGSS*) covered by the experiences in this chapter, as well as links with the *Common Core State Standards*. This chapter only deals with the analysis of data as it relates to key phenomena on the Sun. It focuses on the sunspot cycle, space weather, and related concepts that are most accessible to middle school students. Only a subset of the concepts in the disciplinary core ideas—those concepts and phenomena related to the Sun, its activity cycle, and its occasional storms—is dealt with.

TABLE 3.1

Recommended teaching time for each experience

Experience	Time
Engage experiences	
3.1. What Do We Think We Know?	**45 minutes**
3.2. Be a Solar Astronomer	**30 minutes**
Explore experiences	
3.3. Safe Solar Viewing: Project and Record Your Own Images of the Sun	**60 minutes** or more depending on how many parts you do
3.4. Discover the Sunspot Cycle	**2 school periods**
Explain experiences	
3.5. How Fast Does the Sun Rotate?	**60 minutes**
3.6. Space Weather: Storms From the Sun	**2 school periods**
3.7. What Else Cycles Like the Sun?	**1 to 2 school periods**
Elaborate experiences	
3.8. The Multicolored Sun: An Introduction to Electromagnetic Radiation	**2 to 3 school periods**
3.9. Student Detectives and the Ultraviolet Sun	**2 school periods**
3.10. Additional Ways of Observing the Sun Safely	**30–60 minutes**, depending on how many parts you do
Evaluate experiences	
3.11. Space Weather Report	**15 minutes** to explain, then independent student work, and **30–50 minutes** for presentations
3.12. Predict the Next Sunspot Maximum and Minimum	**30 minutes**

TABLE 3.2

Chapter 3 *Next Generation Science Standards* and *Common Core State Standards* connections

Performance expectations	• MS-ESS1-3: Analyze and interpret data to determine scale properties of objects in the solar system. … Emphasis is on the analysis of data from Earth-based instruments, space-based telescopes, and spacecraft to determine similarities and differences among solar system objects.
	• HS–ESS1-1: Develop a model based on evidence to illustrate the life span of the Sun and the role of nuclear fusion in the Sun's core to release energy that eventually reaches Earth in the form of radiation. Examples of evidence for the model include observations of … the way that the Sun's radiation varies due to sudden solar flares ("space weather"), the 11-year sunspot cycle, and non-cyclic variations over centuries.
Disciplinary core ideas	• MS-ESS1.B: (Earth and the Solar System): The solar system consists of the Sun and a collection of objects, including planets, their moons, and asteroids that are held in orbit around the Sun by its gravitational pull on them.
	• HS-ESS1.A: (The Universe and Its Stars): The star called the Sun is changing and will burn out over a lifespan of approximately 10 billion years.
Science and engineering practices	• Use, synthesize, and develop models to predict and show relationships among variables between systems and their components (e.g., developing an understanding of the relationship of the sunspot cycle and other phenomena on the Sun and Earth; using solar images with visible sunspots to measure their movement and calculate the rotation rate of the Sun).
	• Analyze quantitative data to distinguish between correlation and causation (e.g., analyzing sunspot cycle data to identify the relationship with Earth-based phenomena.)
Crosscutting concepts	• Patterns can be used to identify cause-and-effect relationships (e.g., understanding the pattern in sunspot formation can lead to understanding the rotation of the Sun and the relationship of sunspot activity with other phenomena on the Sun and Earth).
	• The significance of a phenomenon is dependent on the scale, proportion, and quantity at which it occurs (e.g., the impact of space weather on the Earth is dependent on its magnitude and direction of its emission).

Table 3.2 (*continued*)

Connections to the *Common Core State Standards*	• *Writing:* Students write arguments that support claims with logical reasoning and relevant evidence, using accurate, credible sources and demonstrating an understanding of the topic or text. The reasons and evidence are logically organized, including the use of visual displays as appropriate.
	• *Speaking and listening:* Students engage effectively in a range of collaborative discussions (one-on-one, in groups, and teacher-led) with diverse partners, building on others' ideas and expressing their own clearly. Report on a topic or text or present an opinion, sequencing ideas logically and using appropriate facts and relevant, descriptive details (including visual displays as appropriate) to support main ideas or themes.
	• *Reading:* Students quote accurately from a text when explaining what the text says and when drawing inferences from the text. Students determine the meaning of general academic and domain-specific words and phrases in a text relevant to the student's grade level.
	• *Mathematics:* Students recognize and use proportional reasoning to solve real-world and mathematical problems. Students summarize numerical data sets in relation to their context, including reporting the number of observations and describing the nature of the attribute under investigation, including how it was measured and its units of measurement.

Content Background

INTRODUCING THE SUN

The Sun, the star at the center of our solar system, is the main source of light and heat for planet Earth and its inhabitants (Figure 3.2). It is about 150 million km (93 million mi.) from Earth, a distance so large that students need an analogy to fully appreciate it. If the Sun were shrunk down to be the size of a bowling ball, the Earth would be 26 m (about 26 yd.) from the ball, and the Earth would be only 2 mm across (the size of a peppercorn)!

Our Sun, like all stars, is so hot that it is made entirely of gas. In fact, it is so hot inside the Sun that electrons are generally stripped away from their atoms, meaning the negative electrons and

FIGURE 3.2

The Sun as seen from Earth

positive atomic nuclei can go their separate ways. Such a super-hot, charged gas is called a *plasma*. The fact that particles coming from the Sun are often charged (positive or negative) explains their interaction with the Earth's magnetic field.

The Sun produces vast quantities of energy in the form of electromagnetic waves (such as visible light, infrared, and UV). These waves travel away from the Sun at the speed of light and transfer energy from our star to the objects in the solar system, such as the Earth.

Since the Sun shines equally in all directions, only a small fraction of the Sun's energy reaches the Earth. But that small fraction still represents a huge reservoir of energy for our planet. Measurements indicate that 1,361 W of power strike every square meter of the Earth's atmosphere on the side facing the Sun. (This number is often called the *solar constant*, although the amount is not exactly constant but can vary a tiny bit with the Sun's activity level.) You need about 40 W of power to light up a modern lightbulb for reading. So the Sun's power falling on just one square meter of the Earth's surface would be enough to light up about 34 such bulbs.

Another way to understand the enormous amount of energy coming from the Sun is to compare the power reaching the Earth's surface (89,000 terawatts, where a terawatt is a trillion watts) to the total power requirements of all human civilization on Earth (18 terawatts). So the Sun's power output could keep 4,900 civilizations like ours going at the same time, if we could only capture it efficiently.

THE LAYERS OF THE SUN

The production of energy happens deep inside the Sun, where temperatures are in the millions of degrees. What we see when we look at the

FIGURE 3.3

The outer layers of the Sun, with temperatures shown

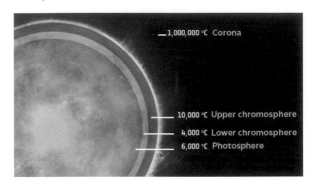

Sun is its outer layer, where the energy emerges (Figure 3.3). We call this the Sun's *photosphere*, meaning the sphere that the light we see comes from. Above the Sun's photosphere lies the Sun's thinner atmosphere, which we generally observe in light other than visible light (e.g., x-rays, UV rays, etc.). During eclipses, however, when the Moon hides the Sun, including the photosphere, some of this atmosphere becomes visible to us through its much fainter visible light. In recent years, an armada of spacecraft has been launched with specialized instruments on board to observe the Sun at many different wavelengths of light and see all the layers without having to wait for special situations like an eclipse.

The layer just above the photosphere is called the *chromosphere* (called by that name since we can sometimes see a flash of color from it around the edge of the Moon when it eclipses the Sun). It is only about 2,000–3,000 km thick but is hotter than the photosphere. Above the chromosphere, temperatures of the Sun's gas continue to increase until they reach millions of degrees in the much wider region called the *corona*.

At first it seemed mysterious that the Sun's *outer* layers should be warmer than its photosphere. Shouldn't temperatures decrease as we get farther from a source of heat? Astronomers eventually realized that some mechanism had to be adding extra heat to these outer layers of the Sun. That mechanism is magnetism, but the magnetic field of the Sun is far more complex than the simple bar magnet we all like to show our students. The surface of the Sun is covered with giant magnetic loops (the shapes of which remind us of the small loops that make up the surface of a carpet). When those loops connect, break, reform, and interact in the Sun's seething hot outer layers, magnetic energy is released, which can heat the gases to even higher temperatures. That same kind of energy can, when the magnetic release is especially powerful, also propel huge lumps of plasma out into space and toward the rest of the solar system.

SUNSPOTS

For many centuries, observers have glimpsed darker areas on the Sun's visible surface and called them sunspots (Figure 3.4). Today we can use filters to cut down the Sun's intensity and see sunspots much more clearly. The typical spot you can see without a telescope is bigger than the entire planet Earth. Some of the biggest spots have been more than 100,000 mi. across—bigger than Jupiter, the largest planet in our solar system! (The entire Earth is about 8,000 mi. across, for comparison.)

The spots last different amounts of time, depending on their size. Smaller ones last only a few hours, while the biggest spots can last months. After a spot disappears from view, other spots often form in the same area of the Sun's surface.

Sunspots look darker because they are cooler than the rest of the Sun's photosphere.

FIGURE 3.4

The Sun with many sunspots (photo taken in 2001 during NASA's SOHO mission)

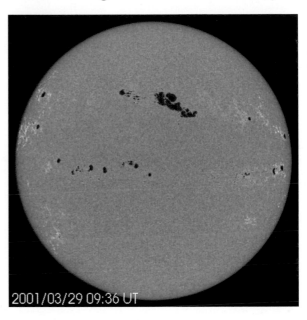

2001/03/29 09:36 UT

Temperatures in the dark sunspots can range from 2700°C to 4200°C (centigrade or Celsius), which is from 4900°F to 7600°F (Fahrenheit). For comparison, the Sun's photosphere is generally at 5500°C or 9900°F. What cools the sunspot areas? The magnetic forces we discussed earlier are so strong in the areas of sunspots that they keep hot material from entering there, allowing the region to cool.

The number of sunspots visible varies over time. Individual spots come and go, but over the years, the overall sunspot counts have shown a remarkably regular pattern of rising and falling. Astronomers have been monitoring and counting sunspots since the 1600s, but it wasn't until 1843 that Samuel Heinrich Schwabe, a German astronomer, was able to show that the number of spots got larger and smaller in a cycle of 11 years.

FIGURE 3.5

August 31, 2012, CME on the Sun, seen with the Solar Dynamics Observatory spacecraft

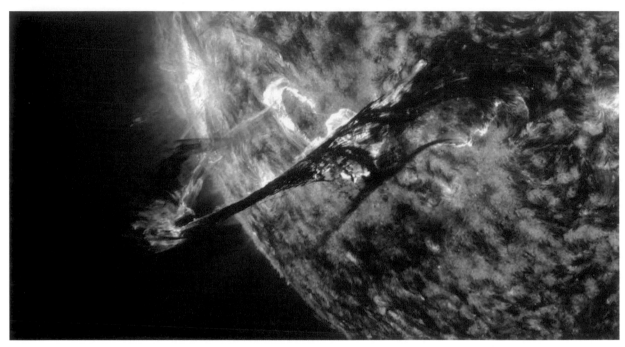

If you want to delve deeper into this with the students, you can discuss how the cycle is not always precisely 11 years. In the past, it has sometimes happened in as short a time as 9 years or taken as long as 14 years, but the average is 11 years.

And we now know that other characteristics of the Sun also change as the number of spots rises and falls. We call these changing properties of the outer layers of our star *solar activity*.

SOLAR ACTIVITY

The time of sunspot maximum is also a time when the Sun is likely to have even more violent expressions of its activity. Among these are flares, sudden releases of visible light, invisible forms of radiation (including x-rays), and charged atomic particles from the seething surface of the Sun; and prominences, tongues or loops of gas from the Sun's surface layer that extend out into its corona.

Perhaps the most dramatic effect a big flare can have is a particularly large blast of high-energy particles from the Sun's outer atmosphere, the CMEs we talked about before. Such a cloud of particles comes off the Sun in one specific direction, and there is no reason why that direction should point toward the location of the Earth at that moment. As a result, most CMEs miss our planet and just move onward in the solar system (Figure 3.5). But once in a while, our planet is unfortunately right in the way of a CME, and then a huge number of high-energy charged particles flow toward our location in

FIGURE 3.6

Diagram showing the Earth's magnetosphere

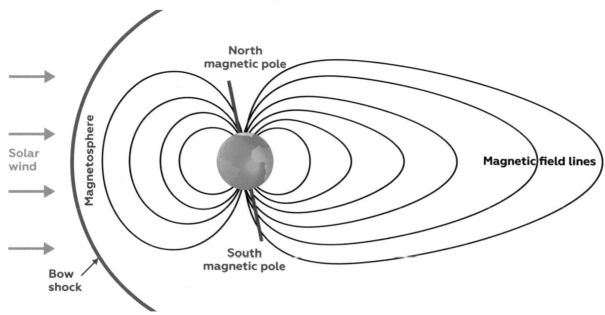

space in a short time. We call that a solar storm when it gets to Earth. (For more, see the Space Weather section below.)

INTERACTING WITH THE EARTH'S MAGNETOSPHERE

The Earth also has magnetic properties: The metals in our spinning core set up a strong magnetic field. In effect, it is as if the Earth had a large bar magnet inside, stretched between the South Pole and the North Pole. The magnetic field from the Earth's "magnet" extends beyond our atmosphere into space. When the charged particles from a flare or a CME reach the Earth, they first meet the magnetic zone around our planet. Called the magnetosphere, this bubble of magnetic influence envelops the Earth. The magnetosphere protects us from charged particles from space, either by deflecting them (moving them aside from the Earth) or by trapping them so that they circle the Earth within the magnetosphere, following its lines of force (Figure 3.6).

Particles trapped in the magnetosphere can't easily enter the Earth's atmosphere at the equator or at lower latitudes. As they follow the lines of our magnetic field, they come down to the ground where the magnetic field lines do—at the magnetic North and South Poles of our planet. When energetic particles come down at those high latitudes, they hit atoms or molecules in our atmosphere and cause them to glow, much like energetic charged particles cause the glow in the neon lights of Broadway. The resulting curtains of light in the sky are called *auroras* or *northern lights* (near the South Pole, they are called *southern lights*).

FIGURE 3.7

Northern lights (aurora) over the Lyngen fjord in March 2012

GEOMAGNETIC ACTIVITY, AURORAS, AND THE K AND A INDEXES

As more charged particles hit the magnetosphere, more of them are trapped in the lines of magnetic force and eventually spiral down toward our planet's poles and hit the atmosphere. At times when flares or CMEs add more particles to the output of the Sun, scientists say that Earth's *geomagnetic activity* is increasing, and tourists in places such as Alaska are delighted by more frequent auroras (Figure 3.7). When a big CME hits our magnetosphere, it can be so overloaded with charged particles that auroras are seen much farther south than usual.

One measure of geomagnetic activity was introduced in its commonly used form by Julius Bartels in 1938 and is called the K index. The larger the K number, which goes from 0 to 9, the more the magnetosphere is being "overloaded" by the output of the Sun and the further south auroras are visible in the Northern Hemisphere. Observatories around the world monitor auroras and other indications of geomagnetic activity and record these local measures to contribute to the K index.

The K index is a compressed (quasi-logarithmic) kind of a number, which is hard to use in averaging. It can be converted into a more stretched A index, which varies over a much wider range than just 0 through 9. We use a version of A index called the aa index in Experience 3.7.

THE SUN'S ROTATION

It was Galileo Galilei who first learned (with help from an associate) to project an image of the Sun from his telescope onto a wall and thus to observe sunspots safely. He was able to satisfy himself that sunspots were on the surface of the Sun (and not, for example, dark objects in orbit around it). This then allowed him to measure how long a big group of spots took to go completely around the Sun and therefore to see how long the Sun took to rotate (see Figure 3.8 for modern images of what Galileo observed).

Since it is made entirely of plasma, the Sun doesn't have to rotate like a solid body. Thus, it shows different rates of rotation on its surface at different latitudes. At the equator, the Sun takes about 24.5 days to rotate relative to the stars. (Because the Earth also moves around the Sun while we are measuring, we on Earth see the Sun's equator take about 26.25 days to rotate.) Near the poles, the Sun takes about 34 days to rotate relative to the stars.

In the 1850s, Richard Carrington, a British amateur astronomer, whom we will meet again in our discussion of space weather, made careful measurements of the rotation rate of the Sun at different latitudes. In honor of his work, astronomers often cite as the Sun's rotation rate the *Carrington rotation period*, measured at a latitude of about 26° on the Sun, the latitude at which sunspots are most likely to form. The Carrington rotation period is about 27.25 days relative to the Earth (and 25.4 days relative to the stars).

When students use sunspots to determine how long the Sun takes to rotate, they are likely to get the Carrington rotation period of about 27 days for their answer.

FIGURE 3.8

Sunspots moving as the Sun rotates, September 27 to October 2, 2011

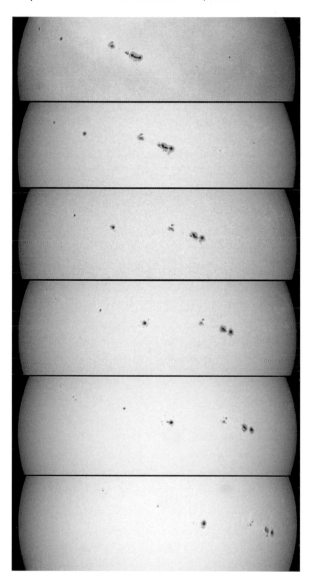

SPACE WEATHER

In addition to the electromagnetic radiation (waves of energy) that the Sun gives off, it also produces a steady "wind" of atomic particles that are launched away from the Sun by its tremendous heat and powerful magnetic energy. This *solar wind* blows outward into the solar system and intersects all the planets, including the Earth.

When strong solar activity increases, the number of energetic particles leaving the Sun, we can call these solar gusts or storms. In this sense, the Sun can produce "weather" in space—which in turn can affect the planets that orbit the Sun. Sometimes, big solar storms can have a powerful effect on Earth, as we discuss below.

On September 1, 1859, amateur astronomer Richard Carrington—whom we mentioned earlier—noticed an especially bright flare on the Sun (it came after a week of increased solar activity). Although Carrington didn't know it (and the concept did not yet exist in astronomy), the flare sent a powerful CME directly at the Earth. To make things worse, there had been an earlier CME that had "cleared a path" in the steady wind of particles from the Sun, and so the second CME found fewer particles in its way between the Sun and Earth than normally and got here with unusual speed and power.

By September 2, 1859, a huge geomagnetic storm had arrived at Earth. Now, this was the time in history when the telegraph, a machine and network for sending coded messages over electrical wires, was spreading all over the United States. Rival companies were offering competing services in different parts of the country, with all of them stringing up wires on telegraph poles for sending messages. As the Earth's magnetosphere was overloaded by charged particles (electricity) from the CME,

dramatic effects were seen all around our planet, but especially in this new electric network.

Auroras were visible the next evening as far south as Hawaii and the Caribbean, and these northern lights were so bright, some people woke up in the middle of the night thinking it was already day. Some people reported being able to read a newspaper in the middle of the night by the light of the auroras. Even more dramatic, the charged particles descending through our atmosphere got into the telegraph system, and some telegraph operators reported sparks coming out of their instruments, while others saw sparks flying from the telegraph wires overhead. The episode is now called the Carrington event in honor of the person who first connected it with activity on the Sun.

In a way, it was lucky that the electrical "grid" in 1859 was still in its infancy. In 2013, a group of scientists and insurance underwriters reviewed records from the Carrington event and tried to estimate the damage this large storm would cause for our far more electrically connected and vulnerable civilization today. They calculated that the damage could be in the trillions (thousands of billions) of dollars!

We had a taste of what can happen in our modern world when a milder geomagnetic storm hit the Earth in March of 1989. Huge currents of charged particles circulated throughout our planet's upper atmosphere. Auroras were visible in both Canada and the United States. Electricity and lights went out in Montreal and other parts of the province of Quebec, leaving Montreal's underground system of pedestrian walkways dark until backup lights could come on. Elsewhere, people reported their electric garage doors opening and closing randomly.

It just so happened that the Space Shuttle Discovery went into space on March 13 that year,

FIGURE 3.9

William Herschel discovered infrared radiation.

just as the storm was arriving, and the astronauts reported some strange readings on their instruments. A number of satellites already in space also reported damage to their electronic instruments because of the surge of electrically charged particles that hit them.

If an even stronger storm from the Sun had reached the Earth, these same types of effects could have been far greater and caused more damage. Although officials at power companies are now starting to take measures to protect the North American power grid (which is quite interconnected) from such damage, such upgrades are very costly and will take time. Until they are made, a really huge solar storm like the

A Little Infrared and UV History

The discovery of waves that are like light but not visible to the human eye began in 1800 when astronomer (and composer and conductor) William Herschel (Figure 3.9) used a prism to spread the light of the Sun into a spectrum of colors—from red to violet. He did an experiment to measure how hot each color of light was and found not only that red was hotter than violet but also that if he moved his thermometer beyond the red, where no light could be seen, it got even hotter. This led him to propose that there was an invisible heating component to sunlight and that it was these invisible waves that led to the hotter temperatures just beyond the red. Today we call those waves infrared radiation (infra means below—as in the word infrastructure—and the frequency of these waves is below that of red light).

A year later, chemist Johann Ritter, inspired by Herschel's work, tried to find something that would reveal invisible waves on the other end of the spectrum, near the color violet. After trying a variety of "detectors" (devices like Herschel's thermometer that would tell him that invisible waves were present), he settled on strips covered with silver chloride, a compound that changes color when exposed to the bluer colors of light. He found that if he put the strips beyond the violet end of the visible spectrum, they changed color even faster. This showed there must be an

A Little Infrared and UV History (*continued*)

invisible wave beyond the violet, just like the invisible waves beyond the red. These were eventually called UV radiation.

You and your students can find historical resources to explore these experiments further here:

- Video dramatizing Herschel's discovery of infrared radiation (2 min.): *www.youtube.com/watch?v=_L7UlqldGuQ*

- A detailed description of Herschel's experiment from the Nuffield Foundation and the Institute of Physics (text): *www.nuffieldfoundation.org/practical-physics/william-herschel-and-discovery-infra-red-radiation*

- How Johann Ritter discovered UV radiation (text): *http://coolcosmos.ipac.caltech.edu/cosmic_classroom/classroom_activities/ritter_bio.html* or *http://io9.com/5948498/how-did-we-discover-the-existence-of-UV-light*

Carrington event could overload the power grid and crash it for our entire continent. Think of all the things that depend on reliable electricity in our world.

There is also some concern about how astronauts in orbit aboard the ISS would be affected by a major CME coming in their direction. So far, no such event has threatened humans inside the ISS. When there is the prediction of a solar storm, astronauts are advised to go to sections of the ISS that offer greater shielding, such as the U.S. Destiny Laboratory or the Russian service module Zvezda.

THE SPECTRUM OF ELECTROMAGNETIC RADIATION (AND THE SUN)

At the temperature of the photosphere, the Sun emits a range (or *spectrum*) of waves. Our Sun's white light is the combination of waves of many colors or wavelengths. Your students may have seen the light from the Sun split into a rainbow of colors when it passes through a prism or diffraction grating.

While a great deal of the Sun's radiation is visible light, there are also waves coming from the Sun that are not visible to the human eye. The Sun gives off a considerable amount of infrared and a smaller amount of UV radiation. These can be detected with specialized instruments even if our eyes cannot see them.

By the way, the word radiation has the same root as radius. Just as the radius of a circle is measured in directions away from the circle's center, so electromagnetic radiation is energy that moves away from where it is produced in all directions (or *radiates* outward). All electromagnetic radiation moves away from its source at the same speed, the fastest possible speed in the universe—the speed of light.

Although the majority of the UV radiation from the Sun is absorbed by the Earth's ozone layer, enough of it gets through that it can cause suntans, sunburns, and even skin cancer (after prolonged exposure), particularly to those not protected by darker skin coloring.

Infrared radiation is often called heat radiation in popular writing because it is the radiation we humans and our environment emit at the temperatures at which we are comfortable. The Earth receives infrared radiation from the Sun and produces its own infrared radiation as the Sun warms it. Calculations of the Earth's energy budget require figuring out all the input and output of infrared waves in and around our planet.

Infrared radiation is absorbed by water, carbon dioxide, and other materials in our atmosphere. As more infrared is absorbed in the air, less of it can escape into space. It is this buildup (due in large measure to human pollution) that is causing *global warming*, the buildup of temperatures under our planet's atmosphere.

The very hot regions of the Sun's atmosphere (in and near the corona) produce x-rays, invisible waves that are even more energetic than UV rays. These do not reach the Earth's surface because they are filtered out in our atmosphere, but they can be observed using x-ray detectors above the Earth's atmosphere. Many of the modern Sun-observing satellites launched by NASA and other space agencies have instruments on board for detecting the invisible waves from our star.

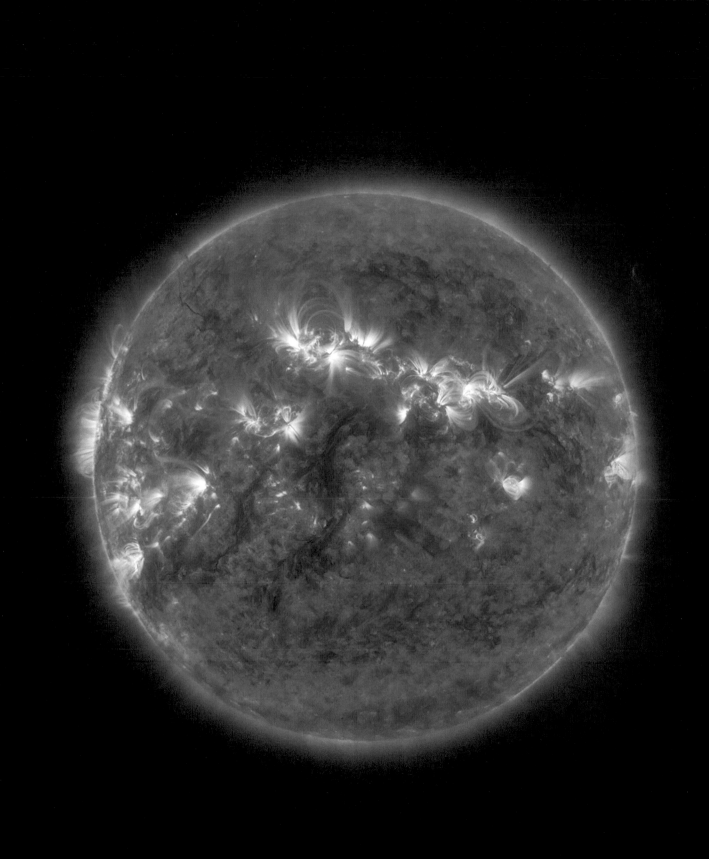

An extreme UV image of
the Sun

EXPERIENCE 3.1

What Do We Think We Know?

Overall Concept

Students will consider and discuss their thoughts, first in small groups and then with the whole class, around the following questions: (1) What features can we observe on the surface of the Sun and in its atmosphere? (2) Does the Sun rotate, and if so, how can we tell?

Objectives

Students will

1. determine their previous knowledge (preconceptions) about evidence of the Sun's activity (such as sunspots or flares) and the Sun's rotation and

2. discuss these two issues among themselves before learning more.

Advance Preparation

1. Identify space in the classroom where student ideas will be posted and can stay up for the entire unit.

2. You may want to have a table of the rotation periods of the planets handy in case students later have questions about how other worlds rotate. (See, for example, *www.bobthealien.co.uk/table.htm*.)

MATERIALS

For the class:

- Space on a whiteboard, blackboard, or piece of poster paper where student ideas can be recorded and kept visible for the duration of this unit

One per student:

- Astronomy lab notebook

Procedure

1. Explain to the students that they will be studying what happens on and around the Sun in this unit and that you want them to think about what they already know about the Sun. First, each student will write down his or her own ideas in his or her astronomy lab notebook, and then the class will have small-group and class-wide discussions.

2. Ask the students to label a new page in their astronomy lab notebooks with "What I Think I Know." Right below that, they should write the first question you want them to consider: "What features can we observe on the surface of the Sun and in its atmosphere?" (If there are questions about what you mean by "feature," you can say that a feature on the Sun can be anything that's smaller than the Sun itself or it can be something that is happening on the Sun.) On the next page of their notebooks, ask them to write the second prompt: "Does the Sun rotate, and if so, how can we tell?"

3. After students write for a while on their own, have them form small groups and pool their ideas. When the groups are done, ask them to share their ideas with the whole class.

4. You or the students can put a summary of the ideas the groups came up with on a whiteboard, blackboard, or poster that can be kept up for the rest of the unit. You can either discuss each question in sequence or have the students think-pair-share about both questions at the same time.

EXPERIENCE 3.2

Be a Solar Astronomer

Overall Concept

A key part of science is that the answers aren't in a book or manual; scientists have to examine things and then find answers to interesting questions. In this open-ended activity, students do what scientists studying the Sun often do—examine a variety of images of the Sun taken over time. They are asked to discuss ideas and questions on the basis of looking at these images.

Objective

Students will notice that changes are visible on the Sun's disk on images of the Sun taken a few days apart, especially in the appearance and number of sunspots.

Advance Preparation

1. Make copies of the handout with the four images of the Sun, making sure the copies are good enough quality to clearly see the sunspots. Since the images are artificially colored, it is fine to copy them in black and white.

2. If you are going to compare these to the current image of the Sun taken with the same space instrument, check that you have a working connection to the internet before class starts.

3. There are many places on the web to obtain current images of the Sun. We recommend one that is relatively easy to use. If you have time, you may want to become familiar with an online application called Helioviewer at *http://helioviewer.org*. This website allows you to bring up an image of the Sun from many different space

MATERIALS

For the class:

* Space on a whiteboard, blackboard, or piece of poster paper to record the ideas that the whole class has shared

One per group of three to four students:

* "Sheet of Solar Images From the Solar Dynamics Observatory" (p. 177), showing visible-light images of the Sun on different days

* Computer with access to the internet if you want to show today's solar image taken with the same instrument that produced the image in Figure 3.10 (p. 176).

One per student:

* Astronomy lab notebook

observatories for any date. Note that you can get to the user's guide by clicking on the box at the upper right or by going to the small help button at the bottom.

Procedure

FIGURE 3.10

Image of sunspots on the Sun

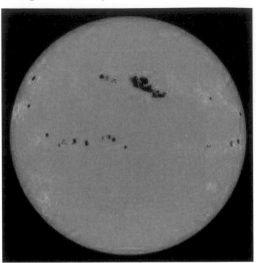

Note that the image has been artificially colored: The Sun is not orange.

1. Organize the students into small groups and give them a copy of the sheet with four visible light images of the Sun from 2014. Ask them to discuss what they notice on the images and what questions they have.

2. If you have access to the internet, show them today's image of the Sun taken with the same instrument. Go to *http://sohowww.nascom.nasa.gov/data/realtime/realtime-update.html* and click on the first image in the second row (the one labeled "SDO/HMI Continuum"). Clicking will enlarge the image. Ask student groups to discuss the differences between today's image and the ones on their sheet. Alternatively, show them Figure 3.10.

3. Have the groups report to the full class and make a list on the whiteboard or a poster paper of all their observations and any questions. (*The students should have noticed that there are dark spots on the face of the Sun and that the number, shape, and position of these spots changes.*) This is not the time to explain fully what sunspots are; instead, you could ask students what might make dark spots on the Sun and record all their suggestions on the whiteboard or poster paper. Students should make notes about their current thinking in their astronomy lab notebooks. These spots will be the subject of Experience 3.4.

Sheet of Solar Images From the Solar Dynamics Observatory

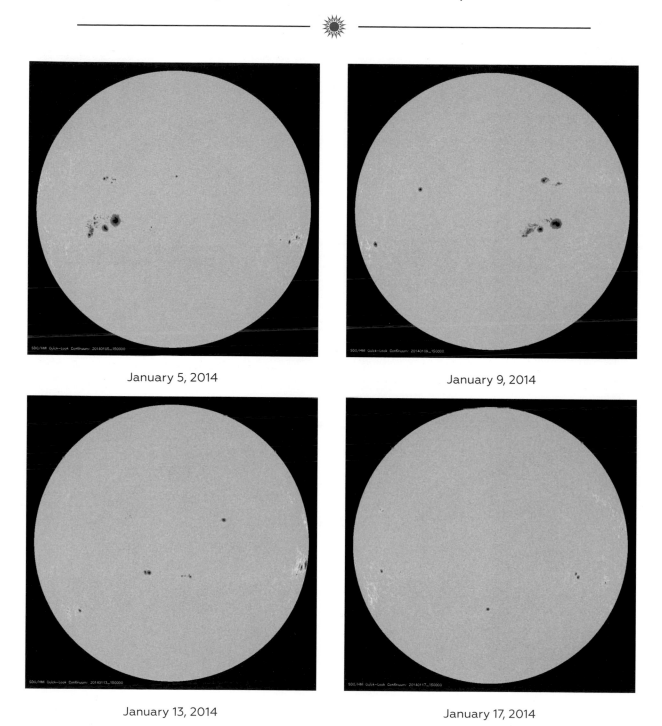

January 5, 2014

January 9, 2014

January 13, 2014

January 17, 2014

Note: If you have color images of the Sun, the color has been added. If you could see the Sun without burning your eyes, it would look generally white in color.

EXPERIENCE 3.3

Safe Solar Viewing

PROJECT AND RECORD YOUR OWN IMAGES OF THE SUN

Overall Concept

This experience provides one safe method, using binoculars, to let students project and examine an image of the Sun, including any sunspots on the Sun's surface.

Objectives

Students will

1. make a Sun projection instrument out of one half of a pair of binoculars (with safety guidance from the teacher);

2. project an image of the Sun onto a white sheet of paper and sketch the appearance of the Sun (including any sunspots and sunspot groups); and

3. come to understand that looking directly at the Sun is dangerous.

Advance Preparation

1. Before going outside, tape the lens cover to the opening of one side of the binoculars. If there is no lens cover, tape a piece of cardboard over the opening so no light can enter that side of the binoculars.

2. Extend the tripod to its full length and attach the binoculars securely to the tripod head. (Some binoculars have a place to screw in a tripod; if not, then tape the binoculars to the tripod head so that it is held tight but the tripod controls allow you to point the binoculars toward the Sun.)

MATERIALS

One per group:

*   Binoculars (a single pair for the whole class can also work)

*   Tripod for each pair of binoculars (with a way of attaching the binoculars)

*   Roll of masking or duct tape

*   Scissors

*   1 ft. x 1 ft. piece of cardboard

*   White sheets of paper on a clipboard or other hard surface

3. Cut a hole in the center of the cardboard the same size as the binocular opening you did not cover. Slip the cardboard over the opening and tape it in place. See Figure 3.11 for how the setup should look when complete.

4. You will find it useful to practice using the equipment before using it with students.

FIGURE 3.11

A diagram of the binocular setup for Experience 3.3

Safety note: Remind students that it is not safe to look directly at the Sun.

Procedure

1. It is important to stress that students *should never look directly at the Sun* through binoculars or with their naked eyes. Permanent eye damage can occur without an immediate feeling of pain.

2. Explain that you (or the students) are going to project an image of the Sun using a pair of binoculars. This will allow them to observe the surface of the Sun without looking directly at it. They will look for small dark areas on the Sun called sunspots. These will be discussed and explained in more detail in the next experience. For now, students just need to know that some sunspots are round, while others have complex patterns as several spots together make a sunspot group.

3. Explain that the sunspots may seem like tiny specks, but most of the ones that we can see are larger than the Earth! Even when a sunspot seems small, it usually only becomes noticeable to the earthbound observer when it is about the size of the Earth.

4. Now take the students outside and help them point the binoculars roughly in the direction of the Sun. *Remember, under no condition should students look directly through the binoculars at the Sun!* Loosen the control knobs on the tripod. It often takes some practice to get the binoculars aimed directly at the Sun. One good indicator of how close you are to having them aligned is to watch the shadow of the binoculars on the ground. The shadow will be as small as possible when the binoculars are aimed at the Sun. When the binoculars are aimed properly, there will be a spot of bright light (the Sun's image) on the ground behind the binoculars.

5. Once you or the students get the bright area of light on the ground, tighten the tripod control knobs. Have a student hold the piece of white paper (on a clipboard, so it is held smooth even if there is wind) about two feet behind the binoculars so that the area of light falls on the paper. The area of light is an image of the Sun, but it may be out of focus. Adjust the focus knob on the binoculars until you get a sharp image of the Sun on the paper. You can also try moving the paper a bit closer or farther. (*Safety note:* Warn students not to put their hands close to where the light comes out of the eyepiece of the binoculars. The concentrated sunlight coming out can give you a little burn if your hand is near the eyepiece.)

    Teacher note: Sometimes you may see only part of the Sun on the paper because the binoculars are not properly aligned with the Sun. If this occurs, carefully adjust the tripod so the full Sun comes into view.

6. Some students may think that the circular image of the Sun they are seeing is not the entire Sun but just a part of the Sun seen through the circular lens of the binoculars. This is more likely to be an issue when there are no sunspots on the Sun. To convince students that the image is not circular because of the lens, you can cover half the lens that's open (making the open lens the shape of a *D*). The image of the Sun will be dimmer because less light is now coming through the lens, but it will still be circular, not D-shaped.

7. As students examine the image of the Sun, they should notice that it is moving across the paper. In a few minutes, the Sun is out of view completely, and you will need to adjust the tripod to bring

the Sun back in to view. Ask students why this is happening. *It's because the Earth is rotating, which is what causes the Sun to appear to move across the sky throughout the day.* You need to realign the binoculars on a regular basis to keep up with the Sun's changing position in the sky. *At first, this may take students a bit of time, but they will get better at it with practice.*

Teacher note: In general, students should get some practice early on gently moving the binoculars to compensate for the Sun's motion in the sky. One student could be moving the binoculars to keep the Sun in view while others sketch an image. (It takes the Sun about two minutes to move a distance across the sky equal to its own diameter.)

8. Once students have the Sun's image on the piece of paper, they should examine the Sun's entire surface carefully to see if they can find any dark spots. If they are not sure whether the spot is on the Sun, in the binoculars, or on the paper, they should lightly tap the binoculars. If the spot moves with the Sun's image, then it is a sunspot.

9. Have students make a drawing of the Sun on the piece of paper, sketching where they can see dark spots on the disk of the Sun. They should also record on the piece of paper the total number of sunspots seen and the date of the observation (and keep the paper for comparing with sketches made on other days).

10. If possible, students repeat their observations with the group daily, using a new piece of paper to record the location and number of spots. Back in the classroom, put the drawings side by side to compare the number and location of the spot and spot groups.

11. Ask students to discuss what they noticed about the drawings in their groups and write in their astronomy lab notebooks about their findings. Did the sunspots change from one day to the next. How did they change? Were there more or fewer sunspots? Did they change size? shape? position? Did the same sunspots move across the surface of the Sun over a period of days?

12. While students continue to make these daily observations, it's a good time to start them on Experience 3.4, "Discover the Sunspot Cycle," which will provide them with more information about sunspots and when there are fewer or more of them.

EXPERIENCE 3.4

Discover the Sunspot Cycle

Overall Concept

Students learn about sunspots, the cooler areas on the visible surface of the Sun. They learn how to count sunspots and sunspot groups on images of the Sun. They then examine records of sunspot number (SSN) counts over 66 years and graph the number of sunspots versus time. This allows them to discover that the number of sunspots varies from day to day and year to year and that sunspots increase and decrease in a roughly 11-year cycle. Optionally, they can mark the birth years of members of their family on the horizontal (time) axis to give the timescale a more personal touch.

Objectives

Students will

1. count sunspots and sunspot groups on a sample photograph of the Sun (just like solar astronomers do) and calculate the SSN;

2. graph the number of sunspots at the end of each quarter for a period of 66 years; and

3. discover that SSNs increase and decrease on a regular cycle, lasting about 11 years.

Advance Preparation

1. If you are doing the optional personalization part, students should get the birth years of significant members of their families, such as siblings, parents, and grandparents.

MATERIALS

One per group of three to four students:

- "Sunspots Worksheet" (p. 189; since color has been added to the image of the Sun, it is fine to copy the sheet in black and white.)

One per student:

- "Graphing Chart A" (p. 190) or graphing paper

- "Sunspots Data Table" (pp. 191–193) showing the SSN at the end of each quarter from 1948 to 2014

- "Sunspots Handout" (pp. 194–195)

- Set of regular pencils and colored pencils for adding date dots

- (*Optional*) The year of birth of each student's family members (siblings, parents, grandparents, etc.)

2. After each group graphs a section of the SSN versus time data, the different plots will have to be pasted together and combined into a single graph. This means you will need to find a place where this large graph can be on display and students can add colored dots below the graph.

3. Prepare copies of the handouts.

Procedure

1. Tell your students that today they get to act like solar astronomers and do research about sunspots. They are especially going to look at how the number of spots on the Sun changes with time. If you have done previous experiences in which students observed sunspots on projected images or images you provided, this is a good time to remind them about those. You can also teach them a little bit more about what sunspots are and what causes them. (See the information in the Content Background section [pp. 163–164] or give out the "Sunspots Handout.")

2. If the students have drawings of the Sun that they made in a previous experience, ask them in their groups to count how many sunspots they saw on one of those drawings. Do the students agree on these counts? If they don't, ask them why that might be. How did the sunspots vary in size and shape?

   Teacher note: Different student drawings of the Sun made on the same day are likely to differ because some spots might be just at the edge of visibility and not all students may record them. Also, sunspots tend to come in groups, and some students may draw the entire group as one bigger spot, while others might draw the individual spots within the group.

3. Next, introduce students to the idea of the SSN, first defined by the Swiss astronomer Johann Rudolf Wolf in 1848. Astronomers found that some sunspots were seen by themselves, while others congregated in groups. Some smaller individual sunspots were hard to see; they were only visible in larger telescopes. The bigger groups, on the other hand, could often be seen even without the aid of a telescope.

Furthermore, counting overlapping sunspots in groups turned out to be harder than counting the individual spots. Different telescopes might lead to different counts, depending on their ability to make out fine detail. Since a typical sunspot group had 10 spots in it, Wolf suggested that astronomers define an official SSN as

SSN = (10 × # of groups) + # of individual spots

Teacher note: Actually, in the formula astronomers use there is also a factor that accounts for the different telescopes and weather conditions at different observatories, but we will leave that out in introducing students to sunspots at this level.

4. Now give students the worksheet with the practice image of the Sun, which shows a good number of sunspots. Divide students into groups of three to four to count groups and spots and calculate the SSN from the image. (The calculations and answers are given in the accompanying box and Figure 3.12.) They should

---

As you can see below, we have circled six groups of sunspots, so the number of groups = 6.

Counting individual spots on the picture is trickier, since some of the spots are quite hard to see. The number of spots inside each circle is shown, but if students didn't get the same counts, don't let them get discouraged. Even professional astronomers do not always agree on the count. It takes some time to get experience counting sunspots, and the pictures used by professionals generally show more details than the ones we see here.

The number of spots (adding up all the numbers next to the circles) is **6 + 1 + 2 + 5 + 12 + 27 = 53**

So the **SSN = (10 × 6) + 53 = 60 + 53 = 113**

Students may point out that this number (113) counts each spot twice (roughly speaking) and they are right about that. The SSN is not a count of spots but a mathematical measure of how many spots are likely to be there, given that some sunspots are hard to spot.

FIGURE 3.12

Sunspots from the image shown to the students (p. 189) are circled

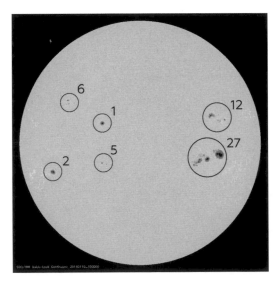

record their data and calculations (using the SSN formula) in their astronomy lab notebooks. Then have the groups share their results for the practice images and discuss any differences in their SSN values. It is not necessary that they all agree on the same value for each photograph. Counting sunspots is a tricky business because some of the spots are quite hard to make out.

5. This is a good time to discuss that the same discrepancies they experienced happen to scientists too. Different scientists might be analyzing the same photographs (or other data) and get different results. Ask students to brainstorm how such differences might be resolved, using SSNs as an example. *Good answers include that different numerical results could be averaged or that observers might practice doing such counts until they got better at estimating what was a spot and what was a group. Experts in sunspot counting could look at a number of photos and write out how they got the SSN in each case so beginners could see how the experts were thinking.*

6. Now that students see how SSNs are calculated, ask them to put away the Sun photo and give each group a copy of the table of SSNs for each quarter from 1948 through 2014 and a copy of "Graphing Chart A." Explain that solar astronomers all over the world have been counting sunspots in this way for many years. They can average their counts over a week, a month, a quarter of the year, or the full year. Have the students examine the table you just gave them. What period of time does each value in the table cover? *One quarter of a year.* Ask them how many years our data chart covers. *It covers 66 years.*

7. Assign each group a period of years to chart, explaining that when they are done, they will put the graphs together to see the complete record. The *x*-axis of their graphs will show the years divided into quarters, while the *y*-axis will show the SSN, which can range from 0 to 250. For each quarter, they should put a dot corresponding to the SSN for the last month in that quarter (so for the first quarter, it's month 3, or March; for the second quarter, it's month 6, or June, etc.).

8. (*Optional*) If the students had the homework of finding out their family's birth years, now ask them if anyone in their

family was born during the years they are working on. If so, the student should put a colored dot at that year on the *x*-axis with his or her initials and how the person is related to them (e.g., mother, aunt, sister). It's helpful to assign a different color of pencil to each group.

9. Now put the graphs from different groups together on a wall or whiteboard, so that the whole 66-year record of SSNs shows. Ask the students individually or in small groups to go up, look at the way SSNs have changed over the years, and search for patterns. They should write in their astronomy lab notebooks about any patterns they see and be as specific as possible. If necessary, encourage them to measure the number of years between times that the number of sunspots was the largest or the smallest. You may need to remind them that each year has four entries, one for each quarter of the year. You may also need to prompt them to look for how much time passes before SSNs go from biggest to smallest and back to biggest (one cycle.) This is a bit tricky, so they should discuss their findings in their groups and then report their estimates to the whole class. Ask them to note their best estimates of when the sunspot minimum and maximum is for each cycle. *If all has gone well, they should be able to identify the roughly 11-year cycle of SSNs (some cycles take less time than this, some take more, but the average is about 11 years).*

Teacher note: If students are new to working with graphs of real scientific data, they may at first be confused because simply connecting the dots by drawing a series of lines from dot to dot doesn't necessarily give them the whole picture. Within the same year, the quarterly SSN may go up and down, but we are interested in the longer-term trend. So, students should try to take in the overall cycle and not get "hung up" on short-term variations. In a sense, they will be visually averaging the points over time, to find the larger-scale trend within the data (Figure 3.13). If they are having trouble with this, you may need to help them out by drawing their attention to the large-scale variations, going from an overall maximum to a minimum and back to maximum. For your reference, a plot of the last few cycles from considerably more data than the students have,

FIGURE 3.13

Sunspot cycle data and predictions

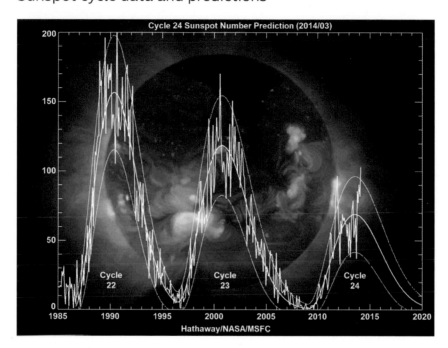

This graph shows the sunspot cycle data compared with the prediction of experts from 1985 to 2015. You can see the current maximum is rather a low one, with fewer sunspots than usual.

and statistically smoothed, can be found at *www.sidc.be/silso/monthlyssnplot*.

10. Mention that although astronomers know that the increase and decrease of sunspots is connected with the magnetism of the Sun and its own cycles of change, the exact mechanism for the sunspot cycle is not fully understood. (This is a good opportunity to mention that there are many parts of astronomy—and science in general—that are not yet fully understood. That's what makes scientists get up in the morning and want to go to work—that they could be contributing to the understanding of things in nature we are still puzzled by.)

11. (*Optional*) The class can now look at the colored dots at the bottom of the graph showing the years that students and their relatives were born. (This doesn't have to be part of the class

work; it could be done during free time by interested students.) For fun, they can see if they and their relatives were born near a sunspot maximum or minimum. When everyone's dots are on the graphs, they should be able to see that there is no connection between birth dates and the Sun's cycle. As many people are born during a minimum or maximum as any other part of the cycle.

Teacher note: Depending on the class, you are likely to find that the colored dots are not randomly distributed. Parents may tend to cluster in a small range of years, as might siblings. The important thing to notice is that there is no correlation with the Sun's 11-year cycle.

12. Now ask students to answer the following questions in their astronomy lab notebooks (referring back to the full graph for the 66 years):

   a. Is the time between the greatest number of sunspots (maximums) always 11 years? How much did the time between the maximums in the graph vary? (Another way to say this is to ask which cycle was the shortest and which was the longest on the whole graph.)

   b. Were all the long-term maximums the same height (i.e., did the sunspot counts rise to the same number in each cycle?)? How much did the maximums vary? For example, what was the largest SSN at sunspot maximum and what was the smallest SSN at sunspot maximum over the 66 years?

   c. What are some other cycles in your life or environment that don't repeat exactly the same way each time? *Cycles of the weather and the seasons should come to mind.*

   d. What was the average number of sunspots in the year you were born? *Take the four quarterly numbers and find the average.* Were you born near sunspot maximum, sunspot minimum, or in between?

   e. What year will you graduate from high school? Will it be a solar maximum, minimum, or in between?

Sunspots Worksheet

Below is a photograph of the Sun taken on January 10, 2014, by an instrument aboard the Solar Dynamics Observatory spacecraft, which has been observing the Sun since 2010.

Please count the number of sunspots and sunspot groups you can see on this image and calculate the sunspot number (SSN) using the following formula:

SSN = (10 × # of groups) + # of individual spots

(If there are sunspots within groups, note that you will, in a sense, be counting them twice. That's OK.)

How many groups of sunspots do you see? _____

How many individual spots do you see (count all)? _____

Use the formula to calculate the SSN:

_____

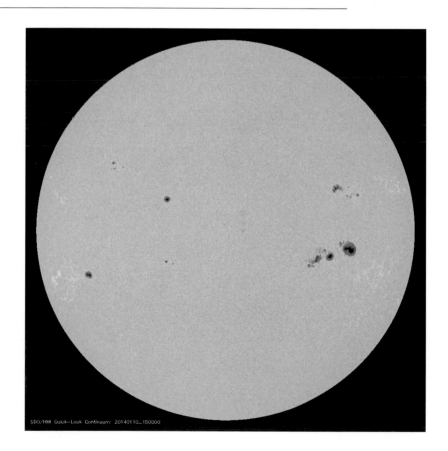

SDO/HMI Quick-Look Continuum 20140110_150000

Graphing Chart A ☀

YEARS

NSS

250 225 200 175 150 125 100 75 50 25

Sunspots Data Table

The table shows the monthly sunspot numbers (at the end of each quarter). The monthly averages (SSNs) were derived from International Sunspot Numbers.

Year	Month 3 SSN	Month 6 SSN	Month 9 SSN	Month 12 SSN
1948	94.8	167.8	143.3	138.0
1949	157.5	121.7	145.3	117.6
1950	109.7	83.6	51.3	54.1
1951	55.9	100.6	83.1	45.8
1952	22.0	36.4	28.2	34.3
1953	10.0	21.8	19.3	2.5
1954	10.9	0.2	1.5	7.6
1955	4.9	31.7	42.7	76.9
1956	118.4	116.6	173.2	192.1
1957	157.4	200.7	235.8	239.4
1958	190.7	171.5	201.2	187.6
1959	185.7	168.7	145.2	125.0
1960	102.2	110.2	127.2	85.6
1961	53.0	77.4	63.6	39.9
1962	45.6	42.0	51.3	23.2
1963	17.1	35.9	38.8	14.9
1964	16.5	9.1	4.7	15.1
1965	11.7	15.9	16.8	17.0
1966	25.3	47.7	50.2	70.4
1967	111.8	67.3	76.8	126.4
1968	92.2	110.3	117.2	109.8
1969	135.8	106.0	91.3	97.9
1970	102.9	106.8	99.5	83.5
1971	60.7	49.8	50.2	82.2

3.4

Sunspots Data Table (*continued*)

Year	Month 3 SSN	Month 6 SSN	Month 9 SSN	Month 12 SSN
1972	80.1	88.0	64.2	45.3
1973	46.0	39.5	59.3	23.3
1974	21.3	36.0	40.2	20.5
1975	11.5	11.4	13.9	7.8
1976	21.9	12.2	13.5	15.3
1977	8.7	38.5	44.0	43.2
1978	76.5	95.1	138.2	122.7
1979	138.0	149.5	188.4	176.3
1980	126.2	157.3	155.0	174.4
1981	135.5	90.9	167.3	150.1
1982	153.8	110.4	118.8	127.0
1983	66.5	91.1	50.3	33.4
1984	83.5	46.1	15.7	18.7
1985	17.2	24.2	3.9	17.3
1986	15.1	1.1	3.8	6.8
1987	14.7	17.4	33.9	27.1
1988	76.2	101.8	120.1	179.2
1989	131.4	196.2	176.7	165.5
1990	140.3	105.4	125.2	129.7
1991	141.9	169.7	125.3	144.4
1992	106.7	65.2	63.9	82.6
1993	69.8	49.8	22.4	48.9
1994	31.7	28.0	25.7	26.2
1995	31.1	15.6	11.8	10.0
1996	9.2	11.8	1.6	13.3
1997	8.7	12.7	51.3	41.2

Page 3

Sunspots Data Table (*continued*)

Year	Month 3 SSN	Month 6 SSN	Month 9 SSN	Month 12 SSN
1998	54.8	70.7	92.9	81.9
1999	68.8	137.7	71.5	84.6
2000	138.5	124.9	109.7	104.4
2001	113.5	134.0	150.7	104.4
2002	98.4	88.3	109.6	80.8
2003	61.1	77.4	48.7	46.5
2004	49.1	43.2	27.7	17.9
2005	24.5	39.3	21.9	41.1
2006	10.6	13.9	14.4	13.6
2007	4.5	12.1	2.4	10.1
2008	9.3	3.4	1.1	0.8
2009	0.7	2.9	4.3	10.8
2010	15.3	13.6	25.2	14.4
2011	55.8	37.0	78.0	73.0
2012	64.3	64.5	61.4	40.8
2013	57.9	52.5	37.0	90.3
2014	91.9	71.0	87.6	N/A

Source: Original numbers courtesy of NASA's Marshall Space Flight Center.

Sunspots Handout

Some History

Galileo Galilei is often given credit for discovering sunspots in 1611, when he pointed his *spyglass* at the Sun. (Today, we call that spyglass a telescope.) We now know that he was not the first person to see sunspots by any means! There are records from ancient Greece and ancient China showing that people almost 2,000 years before Galileo had noticed dark spots on the Sun without a telescope but didn't know what they were.

Galileo looked at the Sun through his telescope only briefly and when the Sun was low in the sky and its light had to go through a large amount of the Earth's air. He could also look when clouds or mist cut down the Sun's light.

However, soon Galileo didn't have to look through his telescope at all to see the Sun. One of his students, Benedetto Castelli, came up with a better idea. He projected an image of the Sun through his telescope at a wall or a big sheet of paper. In this way, Galileo became one of the first people to observe the Sun over a long period of time using a telescope. He made drawings of the spots, which was necessary because the camera had not yet been invented to take photographs (see figure). (Others who did this at about the same time as Galileo include Christoph Scheiner in Germany and Thomas Harriot in England.)

All three observers found that the spots—whatever they were—moved across the face of the Sun as the days went on. Groups of spots would move out of sight on one edge of the Sun and, days later, would sometimes reappear on the other edge. Galileo also suggested that the spots must be on the Sun itself, that they were something that happened in its outer layer or atmosphere. If so, this showed that the Sun was rotating, that is, spinning on its axis just like the Earth. Galileo's idea turned out to be correct.

> **Safety note:** Looking directly at the Sun is dangerous to your eyes, so don't try viewing the Sun without your teacher's guidance. A telescope or a pair of binoculars makes everything brighter, so a view of the Sun through them can damage your eyes even more quickly.

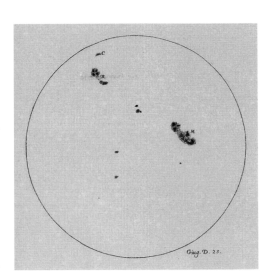

One of Galileo's drawings of the Sun showing sunspots

What Are the Spots?

Today we understand that sunspots are darker areas on the Sun's visible surface (see figure below). Like everything about the Sun, sunspots are huge compared with human sizes. The typical spot you can see with your eyes is bigger than the entire planet Earth. Some of the biggest spots we've seen were more than 100,000 mi. across—a size so big, it boggles our imaginations. The Earth is about 8,000 mi. across, for comparison, and Jupiter, the largest planet in our solar system, is 87,000 mi. across.

The spots last different amounts of time, depending on their size. Smaller ones last only a few hours, while the biggest spots can last for months. After a spot disappears from view, other spots often form in the same neighborhood of the Sun's surface.

Sunspots look darker because they are cooler than the rest of the Sun's outer layer (called the *photosphere*, meaning the sphere from which the Sun's visible light emerges). Temperatures in the dark sunspots can range from 2700°C to 4200°C, which is from 4900°F to 7600°F. The Sun's photosphere, on the other hand, is generally at 5500°C or 9900°F.

(By the way, those numbers are also pretty mind boggling. Water boils at 100°C or 212°F. When we are talking about the Sun's temperatures, it's good to bear in mind that things are so hot there that our bodies would not only boil but also evaporate. The heat would tear our bodies apart until we were just individual atoms of gas, by which time we would long be dead. The Sun is made entirely of superheated gases—so hot that the electrons are separated from their parent atoms.)

It's important to remember that the spots are only dark compared with how super-bright the rest of the Sun is. If, somehow, we could remove the sunspot regions from the Sun, they would glow a rich red and be bright with light.

What cools the sunspot areas? Twentieth century astronomers discovered that the Sun, filled as it is with negative electrons and positive atoms, is electric and magnetic in complicated ways. Astronomers now know that magnetic forces are so strong in the areas of sunspots that they keep hot material from flowing up from below, allowing the region of the spot to cool.

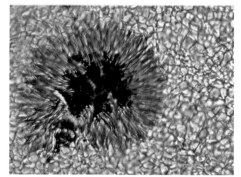

A close-up of a sunspot, taken with a large solar telescope

EXPERIENCE 3.5

How Fast Does the Sun Rotate?

Overall Concept

Students follow the movement of sunspots across the disk of the Sun on satellite images to determine how fast the Sun rotates.

Objectives

Students will

1. learn about measuring location on a round world (latitude and longitude systems) and

2. use the angular displacement of sunspots over a period of seven days to measure the rotation period of the Sun.

Advance Preparation

1. Prepare your demonstration ball by using the erasable marker to draw longitude lines at approximately 15° intervals. (That means 24 evenly spaced lines around the ball, going from the North Pole to South Pole.) Some people first wrap the ball in saran wrap to make removing the lines later less of a chore.

2. Using the marker, put several good-sized sunspots in the midnorthern latitudes.

3. Copy the two sets of solar images (sheet 1 and sheet 2) so that roughly equal numbers of student groups are working with each different set. Although the images are in color in our book, that color has been artificially added

MATERIALS

For the class:

*   1 medium-sized ball (such as a white soccer ball, a basketball, or a beach ball)

*   (*Optional*) Saran or other plastic wrap

*   Erasable markers

*   1 small paper plate made to look like a circular sunspot with a black center and lighter ring around the center

For the groups (each sheet is given to half of the groups):

*   "How Fast Does the Sun Rotate? Sheet 1: Images of the Sun From the Solar Dynamics Observatory in May 2014" (p. 204)

*   "How Fast Does the Sun Rotate? Sheet 2: Images of the Sun From the Solar Dynamics Observatory in November 2011" (p. 205)

*   Transparency overlays with a grid of solar longitude measuring lines (Stonyhurst disks)

by the project scientists: The Sun would actually look white to you if you could look at it safely. So, it is fine if you photocopy the images in black and white.

4. Make transparencies of the Stonyhurst disk master and have one for each group. It's important that the disk on the transparency be the same size as the disk of the Sun on the photographs you are using to track sunspots. In this book, we have made the pictures on sheets 1 and 2 and on the transparency master the same size.

Procedure

1. If students did Experience 3.1, "What Do We Think We Know?" ask them to recall their thinking about the second prompt in that experience: "Does the Sun rotate, and if so, how can we tell?"

   If not, then ask students how we can determine whether the Sun rotates, and if it does, how we can figure out how long it takes to turn. These questions can be answered in small groups (which then share their answers with the whole class) or in a whole-class discussion. Someone in the class will likely suggest that we can follow sunspots until we see them make a complete circle around the Sun. If not, guide them in the direction of this suggestion and then tell them that they are going to follow a number of different sunspot groups to figure out how much time it takes the Sun to rotate.

   Teacher note: This could be a good time to mention to the students that about 400 years ago, Galileo Galilei (using his new spyglass, otherwise known as a telescope) watched groups of sunspots carefully as they moved across the face of the Sun, disappeared on one edge, and later reappeared on the other edge. His observation of sunspots and their shapes eventually led Galileo to conclude that sunspots were features on the surface of the Sun (and not, for example, objects in orbit around the Sun). This meant they could be used to show that the Sun rotated. Furthermore, they allowed him to demonstrate that the Sun was a sphere, something not everyone believed at the time.

2. Use the ball on which you have drawn one or more big dark spots (big enough so the class can see them). Turn it slowly and uniformly with the students watching. Ask them to keep an eye

on what happens when spots get close to the edge of the ball and are about to rotate to the ball's back side. Do this several times as students watch. They should notice that the spots foreshorten (see more on foreshortening in step 3) as they get to the edge of the ball. Also, students may be able to see that the spots appear to move faster near the center of the ball but then more slowly when they get to the edge.

3. This is a good time to explain that the spots appear to move more quickly at the center because all the motion there is lateral (sideways.) Near the edge of a three-dimensional ball, more of the motion is front to back, so the spot appears to move more slowly. This is a subtle idea and may take a while for the students to understand. Being near the edge of the Sun also means that the sunspot looks shorter from front to back as it wraps itself around the curvature of the Sun (but it is not smaller from top to bottom). This is what we call foreshortened (Figure 3.14). You can use the small paper plate made to look like a sunspot (and attached to the rotating ball) to demonstrate this effect. You can also point out to the students that, for the same reason, the longitude lines on the ball appear closer together near the edge of the ball than in the center.

4. Tell the students that they are going to use the same process that Galileo used to measure how long the Sun takes to rotate. They will observe one set of spots over a period of time and measure how long the spots take to move across the Sun. From this, they will be able to figure out the time it takes the Sun to make one rotation.

   Teacher note: The Sun is not a solid body, and thus its parts don't have to "stick together" and move at the same pace. Sunspot groups and individual spots can change and even disappear as students

FIGURE 3.14

Sunspots appear foreshortened near the edge of the Sun.

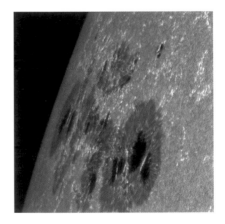

This image, taken November 2, 2011, from the Solar Dynamics Observatory satellite, shows foreshortened sunspots near the left edge of the Sun.

watch them move on the Sun, which can be a source of confusion. Generally, the smaller the spot, the sooner it tends to disappear.

Teacher note: A dramatic one-minute animated movie showing a single large and complicated sunspot group moving and changing across the face of the Sun (from one side to the other) can be found at *http://apod.nasa.gov/apod/ap150629.html*. This was taken with NASA's Solar Dynamics Observatory. You can decide at what point in the experience you want to show this short film.

5. Before introducing the idea of the latitude and longitude of sunspots on the Sun, be sure that students are familiar with latitude and longitude as a system of locating things on the Earth.

Teacher note: If students are not sufficiently familiar with the latitude and longitude circles on Earth, you will need to remind them that latitude circles go around the Earth east–west; the system of circles starts at the equator (0° latitude) and extends north and south from there until it reaches 90° N at the North Pole and 90° S at the South Pole. Longitude circles, on the other hand, go around the Earth north–south; they begin with 0° on a circle connecting the poles with Greenwich, England (this circle is called the prime meridian). The system of circles goes around the globe to +180° as you go eastward and −180° as you go westward. So your latitude tells you your location north or south of the equator, and your longitude tells you your position east or west of the prime meridian. Specifying both latitude and longitude means you have fixed a specific point on the surface of the Earth.

Help students notice that you measure latitude relative to the equator and poles (which are easy to find for a spinning globe) but that you need some landmark or agreed upon starting place from which to specify longitude. (On Earth, there was a fierce struggle between countries about which European city would get the honor of being called the prime meridian. Greenwich, England, won!)

If examples are needed, have students use the internet to find the latitude and longitude of some favorite locations around the Earth (e.g., their home town, Honolulu in Hawaii, the pyramids in Egypt.) For example, they can use the NASA site *http://mynasadata.larc.nasa.gov/latitudelongitude-finder*.

Point out to the students that we can set up an equivalent system to specify locations on the Sun. We can set up a system of

FIGURE 3.15

An illustration of a solar disk, with latitude and longitude lines shown and labeled with degrees

FIGURE 3.16

Stonyhurst disk grid overlay

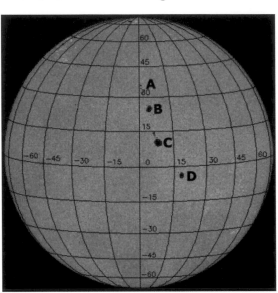

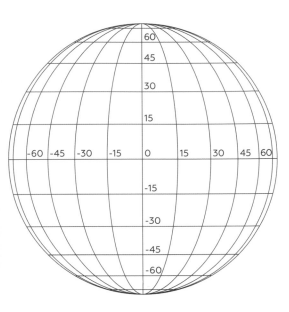

latitude circles on the spinning Sun, just like on Earth, by using the equator and the poles; however, longitude is harder.

Since sunspots come and go and the Sun has no fixed markings on its surface (photosphere), we can be arbitrary about defining longitude for the Sun. For ease in measurement, we will take the 0° longitude to be the vertical line in the middle of the Sun's disk when we first measure the location of a sunspot or sunspot group (Figure 3.15) .

6. Distribute a Stonyhurst disk grid overlay (Figure 3.16; named after the observatory in New York State where they were first used) to each group, together with the images of the Sun that they will be analyzing (sheet 1 or sheet 2). Explain that the grids on these transparencies show a system of latitude and longitude for the Sun, which we can use to see how sunspots are moving on the Sun.

7. Students should assign a letter (A, B, C, etc.) to each sunspot group. They should line up the transparency so that the overlay fits exactly over the solar image and use it to estimate each

TABLE 3.3

Data table for recording sunspot locations

Date	Spot (group) letter	Latitude and longitude	Description of spot or group
	A		
	B		
	C		

spot's latitude and longitude (the angle east or west of the prime meridian [the line running straight up and down in the middle of the template]). Note that all the images of the Sun are taken at the same time on each of the days.

8. For each image, students should make a data table in their astronomy lab notebooks that looks like Table 3.3 (above):

    Students can add more rows if they see more than three large sunspots or sunspot groups on their images. The description column in the table is so that students can keep track of each sunspot or sunspot group as the days go on. It can be about the spot or group's size or shape or darkness. (Remind students that the larger the group of spots, the more stable it tends to be. Smaller spots can disappear in a day, so the larger the group they are working with, the more likely it will last.)

9. For each sunspot or group that they can follow, students should calculate the angular speed of the spot or group using the following formula:

$$\text{angular speed (in degrees per day)} = \frac{\text{(change in longitude in degrees)}}{\text{(number of days between measurements)}}$$

    So, for example, if in two days, a group has changed its longitude by 26°, the angular speed = 26°/2 = 13°/day.

    Teacher note: We are only considering change in longitude in the formula because sunspot groups tend to stay at the latitude where they emerge as they go around. That's something the students may discover for themselves as they look at the pictures, so you may want to hold back that information.

Teacher note: Several issues will come up as students make their measurements. The main one is what location in a sunspot group should be used to mark its location (e.g., the leading edge of the group, the middle, or the center of the largest spot?). This becomes even more of a problem if the sunspot group has changed size or shape between measurements. Reassure the students that solar astronomers doing this kind of work have to deal with this problem all the time. There is no perfect solution to being sure you have the right place to measure each time. This is why students should make measurements for several sunspot groups and get an average (as the instructions suggest). Also, there is the question of whether the images we are using were each taken at the same time of day. If they weren't, then the denominator in the formula might have to include nonintegers, such as 1.2 days. In fact, the images on sheets 1 and 2 were all taken at the same time of day.

10. Students should first measure the angular speed three days apart and then measure it for the entire six days. They can then average its angular speed over the full period to help eliminate the problems encountered because the positions of spots are hard to estimate precisely. They can also measure the angular speed of other spots and groups and compare their values for each.

11. Students now use their values of the average angular speed to calculate how long it takes the spots to go completely around the Sun—thus finding the rotation period of the Sun. If necessary, remind them that a full circle has 360° in it.

rotation period = (360°)/(angular speed in degrees / day)

So, for example, if the angular speed is 13° per day, then the period would be

360°/(13°/day) = 28 days

Each group should derive its own value of the rotation period and then students can come together as the whole class and share their answers, explaining how they calculated them. Have them suggest possible reasons why the groups reached different rotation periods.

Once the class has exhausted their possible explanations, you can sum up. In addition to the issues discussed above (spots changing with time, deciding just where to measure a spot or group, etc.) you may want to share with them that, unlike the Earth, which rotates

202 **NATIONAL SCIENCE TEACHERS ASSOCIATION**

as a solid body and thus has one rotation period, no matter where (at what latitude) you measure, the Sun is a ball of hot, electrically charged gas and its different layers don't have to rotate with the same period. Therefore, the answer students calculate will also depend on the latitude of the spot or group they are measuring.

Teacher note: At the equator, the Sun takes about 24.5 days to rotate relative to the stars. (Because the Earth in the meantime moves around the Sun, we on Earth see the Sun's equator take about 26.25 days to rotate once.) Near the poles, the Sun takes about 34 days to rotate relative to the stars. Such differential rotation is also seen on the large planets made mostly of gas and liquid, such as Jupiter. In the 1850s, Richard Carrington, a British amateur astronomer, made careful measurements of the rotation rate of the Sun at different latitudes. In honor of his work, astronomers often cite as the Sun's rotation period the Carrington rotation period, measured at a latitude of about 26° on the Sun, the latitude where sunspots are most likely to form. The Carrington rotation period is about 27.25 days relative to the Earth (and 25.4 days relative to the stars).

A QuickTime movie showing the Sun rotating over a period of 36 days can be found at *http://solarscience.msfc.nasa.gov/ images/gongmag4.mpg.* (Note that it shows the magnetic intensity of the spots rather than a visible light photo, but it does show the rotation as seen via sunspots quite clearly.)

A more complex film showing the rotating Sun for January 2014 with views at many different wavelengths can be seen at *http://apod.nasa.gov/apod/ap140312.html.* (This is perhaps better to show after Experience 3.11, "The Multicolored Sun.")

For More Advanced Classes

The Sun appears to wobble a bit as seen from Earth over the course of a year because the plane of our orbit is not exactly in the Sun's equatorial plane. To correct for this small effect, you can actually use a different Stonyhurst disk overlay for each of the 12 months, given that the Sun is tilted differently relative to Earth in each month.

To obtain these disks and see this more advanced version of the activity, go to Space Weather Forecast, a curriculum and activity sequence developed at Chabot Space and Science Center for the Stanford Solar Center: *http://solar-center.stanford.edu/SID/educators/ SpaceWeatherForecast-v.070507.pdf.*

How Fast Does the Sun Rotate?

SHEET 1: IMAGES OF THE SUN FROM THE SOLAR DYNAMICS OBSERVATORY IN MAY 2014

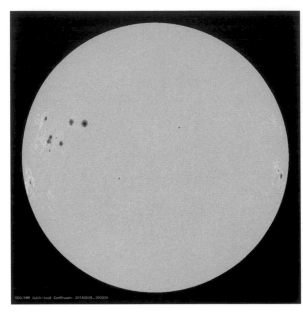

May 8, 2014

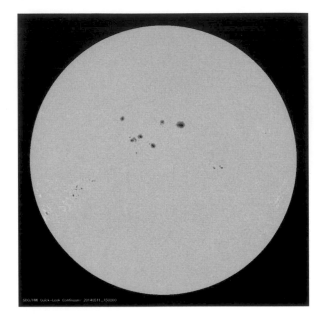

May 11, 2014

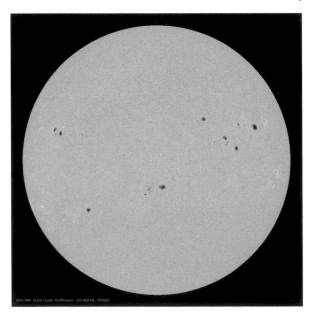

May 14, 2014

How Fast Does the Sun Rotate?

SHEET 2: IMAGES OF THE SUN FROM THE SOLAR DYNAMICS OBSERVATORY IN NOVEMBER 2011

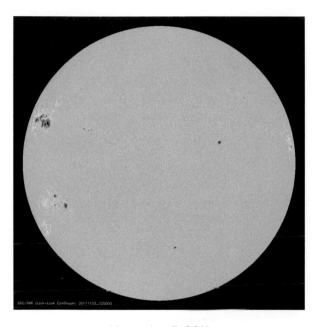

November 3, 2011

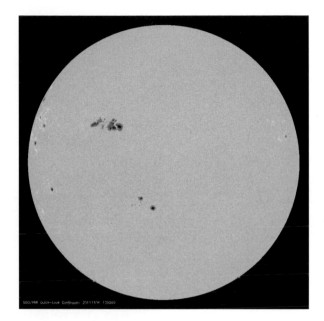

November 6, 2011

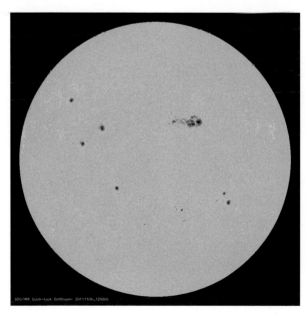

November 9, 2011

EXPERIENCE 3.6

Space Weather

STORMS FROM THE SUN

Overall Concept

Students first brainstorm and discuss what goes into weather reports on Earth (e.g., temperature, wind, gusts of rain or snow, etc.) and how weather reports (online, on TV, in newspapers) are put together (Figure 3.17a). Next, they learn a bit about the concept of space weather and its causes (Figure 3.17b). Then, working in small groups, they make a list of solar events that could eventually affect the Earth. They discuss their thoughts and questions with the entire class. Either in small groups or individually at home, they search for websites with accessible reports on space weather. (If

FIGURE 3.17

Storms on Earth and the Sun compared

 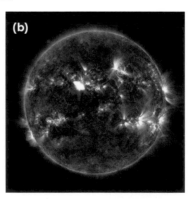

A photo of Hurricane Frances taken by astronaut Mike Fincke aboard the International Space Station from 230 mi. above the storm August 27, 2004. (b) A bright flare is seen near the center of the Sun in this false-color extreme ultraviolet view of our star taken with the Solar Dynamics Observatory satellite in April 2013. The flare launched a CME toward the Earth.

MATERIALS

For the class:

- Space on a whiteboard, blackboard, or piece of poster paper where student ideas and questions can be recorded

- Computers with access to the internet

One per student:

- "Solar Storm Warning and Protection Worksheet" (pp. 211–212)

- "Space Weather Handout" (pp. 213–216)

- Astronomy lab notebook

needed, a list of useful websites is also provided.) In a class discussion, groups compare the results of their research. Students watch a NASA video on a coronal mass ejection (CME) that missed the Earth and read about the Carrington event and other times when a CME did hit the Earth. Then, in small groups, they brainstorm again about the effects that a solar storm could have on the Earth (and their own lives). The class comes together in a symposium to discuss all their ideas.

Objectives

Students will

1. brainstorm about weather on Earth and what goes into weather reports we read or see on TV,

2. learn and brainstorm about the Sun's activity and what part of it might produce space weather,

3. do web research and watch a video about storms from the Sun,

4. brainstorm what effects a storm from the Sun might have on their own lives and on the Earth in general, and

5. participate in a class discussion of these issues.

Advance Preparation

1. Make a copy of the "Solar Storm Warning and Protection Worksheet" and the "Space Weather Handout" for each student.

2. If you are in a computer lab, you may want to bookmark the list of best space weather sites for the level of your students (see Procedure).

Procedure

1. Tell the students that you want them to learn about weather in space and that, before they think about space weather, you want them first to think about weather on Earth. In their astronomy lab notebooks, ask them to write the following question on a new page: "What sorts of factors do weather reports on Earth consist of?" Ask them to fill in as many different factors as they

can think of. After students have written for a while, use the think-pair-share method by having students gather in small groups, compare their ideas, and develop a list for the group to share with the class. Then the different groups can share their lists with the whole class. List their ideas on the whiteboard or a sheet of poster paper.

Teacher note: Factors students might list include the following:

1. Temperature (current, highest, lowest)

2. Rain, drizzle, snow, fog, hail

3. Whether it is completely overcast, cloudy, or sunny (and types of clouds)

4. Wind speed and direction

5. Storm warnings for thunderstorms, tornadoes, cyclones, snowstorms, and blizzards

6. Air pressure

7. Pollen count and other air quality factors.

2. Have students check some weather websites and watch weather reports on local TV news to see how many of their ideas are part of weather reports and what other factors go into weather that they didn't think of. Among weather websites they could check are the following:

    Weather Underground: *www.wunderground.com*
    National Weather Service: *www.weather.gov*
    The Weather Channel: *www.weather.com*

3. Next, students need to learn about weather from the Sun. It's up to you how you want them to learn this information. You could distribute the "Space Weather Handout" the evening before for them to read, use the information in the chapter's contents background section to teach about the topic yourself, or give the students other resources for learning on their own (see the Literacy Connections section at the end of this chapter).

4. Ask students in their groups to talk about space weather in the "neighborhood" of the Sun—a neighborhood that stretches all the

way to the Earth. They should focus on these questions: How can conditions on the Sun create some kind of weather in space, and how could that weather have an effect on Earth? What factors would make the weather in space change from day to day or month to month? They should write ideas in their astronomy lab notebooks. They can also make a list of questions they have about this topic.

Teacher note: This may at first be a difficult topic for students. Reassure them that it's OK not to know a lot about space weather at this stage. Guide them with some hints if needed. (See the Content Background section [pp. 168–170] for ideas.)

5. Once groups have some space weather ideas and questions to share, bring the class together in a "space weather symposium" and have them discuss what they have come up with. List their ideas and questions on the whiteboard or poster paper.

Teacher note: Among examples of space weather that students might discuss are the following:

1. Solar wind (the regular flow of particles from the Sun)

2. Solar wind gusts

3. Flares on the Sun

4. CMEs from the Sun

5. Auroras in the Earth's atmosphere

6. Other examples of increased disturbances in the Earth's magnetic zone.

Factors contributing to the variability of space weather could include the spinning of the Sun, differences in temperature among various parts of the Sun, changing sources of energy or heat inside the Sun, the fact that the Sun is made of hot gas, and so on.

6. Have students watch the following NASA video about the 2012 CME that just missed planet Earth: *www.youtube.com/watch?v=7ukQhycKOFw.*

7. Now distribute the "Solar Storm Warning and Protection Worksheet." In their small groups, students should go through the issues on this sheet and discuss what effects space weather might have on them and on the Earth in general. Then they should brainstorm how we Earthlings might protect ourselves from these effects. They should also list any questions they have.

8. Bring the groups together and have another class discussion, this time focusing on the effect of space storms on Earth and how we might protect our technology. There may be issues that come up during this discussion to which neither you nor the students have an answer. Issues may include exactly how power companies can protect the electrical power grid, how much shielding a satellite in space needs so its computers can keep working during a big storm from the Sun, or how big a storm needs to be before the astronauts aboard the space station get exposed to dangerous radiation levels no matter where in the station they position themselves. Explain to the students that such excellent questions are at the forefront of science and engineering today, and lots of scientists and engineers are thinking about them just like the students are. If students want to know about current thinking in these areas, they can do a library or a web search and perhaps report back to the class for extra credit (see some of the resources at the end of the chapter).

9. Students write in their astronomy lab notebooks about what they have learned about space weather and protecting the Earth from future storms on the Sun.

Solar Storm Warning and Protection Worksheet

Earth's magnetosphere during a solar storm

A diagram showing a storm on the Sun and the magnetosphere around the Earth. As the charged particles leave the Sun and arrive at Earth, many are pushed aside by our planet's magnetosphere. But some spiral inward and collide with the Earth at the poles. This drawing is NOT to scale; the Sun is much farther away from the Earth than this shows.

1. Imagine that you are one of the Space Weather Prediction Center forecasters and that you have just seen a big coronal mass ejection from the Sun. It looks like it's heading in the direction of planet Earth. From its speed, you and your colleagues estimate it will reach Earth in 2.5 days.

    Make a list of the different kinds of organizations and people you will want to notify about the coming solar storm:

    1. _____

    2. _____

    3. _____

    4. _____

    5. _____

    6. _____

    7. _____

    8. _____

    9. _____

    10. _____

Solar Storm Warning and Protection Worksheet

✴

2. Now select four of the organizations or types of people you listed, and discuss what they might do in those 2.5 days to protect against the coming solar storm.

    1. _____
       _____

    2. _____
       _____

    3. _____
       _____

    4. _____
       _____

3. What questions did you and your group members come up with about the above two issues?

    1. _____
       _____

    2. _____
       _____

    3. _____
       _____

    4. _____
       _____

    5. _____
       _____

    6. _____
       _____

Space Weather Handout

Although most of us think of weather as something that happens in the Earth's atmosphere, it turns out that activity on the surface and in the atmosphere of our Sun can also cause its own dramatic kind of weather.

Solar Wind

Because the outer layers of the Sun are so hot, atoms and parts of atoms leave the Sun at speeds that average almost a million miles per hour. (That sounds outrageously fast if a car were driving at that speed, but it's easier to get tiny protons and electrons moving that fast.) We call this flow of material the *solar wind*. As it moves through the whole solar system, some of it streams toward the Earth. Most of the solar wind is made of charged particles: electrons, protons, and alpha particles (the nuclei of helium atoms).

When the charged particles reach our planet, they are caught up in the magnetic zone around the Earth and spiral around. This *magnetosphere* (the zone of our magnetic field) around Earth keeps most of the charged particles from reaching us (see figure below). The only place the magnetic field can lead particles from the solar wind toward Earth is at our north and south magnetic poles. There, the magnetic field comes in and goes out of our planet, and the Sun's fast moving particles can hit the top of our layer of air.

When the solar wind particles collide with air molecules, they cause them to "jiggle" (vibrate) and give off light. Tourists like to go to the northern and southern regions of the Earth and watch the resulting glowing curtains of light high in the atmosphere. We call this an *aurora* (or the northern lights and southern lights).

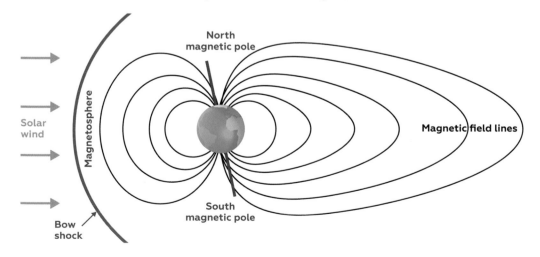

A diagram of Earth's magnetic field

Solar Storms

In addition to its steady wind, the Sun can also have "storms" in its upper layers. These storms are the result of the sudden release of a lot of magnetic energy, and they can fling huge numbers of particles into space at once. Such storms are more likely to happen during times when the Sun has more sunspots on its surface, a time known as solar maximum.

There are a number of ways in which we see increased activity (storms) on the Sun. When there is a flare, a small part of the Sun gets much brighter as both energy and particles are released. In a *coronal mass ejection* (CME), huge bubbles of charged particles are flung from the Sun in one particular direction. A large CME can contain billions of tons of charged particles and take a few days to travel the distance from the Sun to the Earth. Luckily, most CMEs don't happen to be going in our direction, and they travel onward through our solar system without affecting us. But every once in a while, a CME turns out to be flung directly toward Earth and then our cosmic neighborhood (and particularly our magnetosphere) is overloaded with all its energetic particles.

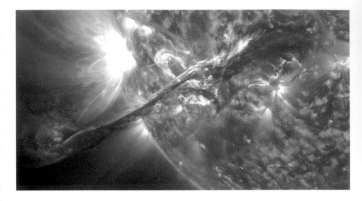

A 2012 CME seen with NASA's Solar Dynamics Observatory spacecraft.
(The colors are enhanced.)

When a storm from the Sun reaches the Earth, auroras can be seen farther south in the Northern Hemisphere and farther north in the Southern Hemisphere. Even more significant, all those extra charged particles set up large currents of electricity in the Earth's upper atmosphere (particularly in the layer called the ionosphere).

Effects on Earth Technology

The extra electricity introduced by a big solar storm can affect many parts of our lives, which are more and more dependent on electricity and delicate instruments like computers. Electricity is supplied to homes and businesses by local and national networks (called the *electric power grid*). Large currents in our atmosphere can overload the power grid and cause electrical blackouts around the local area. Because the grid connects many regions together, such blackouts can sometimes also affect other parts of the country and the continent.

Space Weather Handout

Up in space, the extra flow of charged particles can penetrate computers and other electronic parts in our orbiting satellites. This can lead to computers giving the wrong instructions to the satellite instruments or even to permanent damage when there is a strong storm. Engineers now shield sensitive satellite instruments with protective coverings so they aren't as easily damaged.

Solar storms can also interfere with long-distance communications on our planet. To reach distant places, we often bounce radio waves off the ionosphere, which can act like a mirror, reflecting waves to other parts of the Earth. But if this mirror is disturbed, communications can't get through. Communications problems are especially dangerous for airplanes, which fly up in the atmosphere and need to keep in touch with flight controllers around the world.

There is also concern about how astronauts in orbit aboard the International Space Station (ISS) would be affected by a major CME coming in their direction. So far, no such event has threatened humans inside the ISS. When there is the prediction of a solar storm, astronauts are advised to go to sections of the ISS that offer greater shielding.

Monitoring Space Weather

Because storms from the Sun can be dangerous for human activities, NASA and the National Oceanic and Atmospheric Administration (NOAA) carefully monitor the Sun and the space weather it causes. NASA has a number of satellites in space to keep regular track of what the Sun is doing. The most recent of these (launched in 2010) is the Solar Dynamics Observatory, which monitors the Sun's output

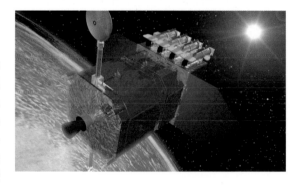

The Solar Dynamics Observatory

not just in visible light, but in ultraviolet light and x-rays, too.

NOAA's Space Weather Prediction Center (in Boulder, Colorado) keeps track of what's happening on the Sun 24 hours a day, seven days a week. If they see conditions that signal a storm coming, they alert military and civilian agencies so that airlines, power grid operators, and others who depend on electricity, satellites, and long-distance communications can take precautions.

For more information, print out the NOAA booklet Space Weather from *www.swpc. noaa.gov/sites/default/files/images/u33/swx_booklet.pdf.*

A History of Solar Storms

On September 1, 1859, amateur astronomer Richard Carrington noticed a bright flare on the Sun. Although Carrington didn't know it then, that flare also came with a powerful CME pointed directly at the Earth. To make things worse, there had been an earlier CME that had "cleared a path" in the steady wind of particles from the Sun, and so the second CME found fewer particles in its way between the Sun and Earth than normal and got here with unusual speed and power.

On September 2, a huge solar storm arrived at Earth. As the Earth's magnetosphere was overloaded by charged particles, dramatic effects were seen all around our planet. Auroras were visible the next evening as far south as Hawaii and the Caribbean, and these northern lights were so bright, some people woke up in the middle of the night thinking it was already day. Some even reported being able to read a newspaper in the middle of the night by their light.

This was the time in history when the telegraph—a machine and network for sending coded messages over electrical wires—was spreading all over the United States. Rival companies were stringing up wires on telegraph poles for sending messages between people and companies. (You could say this was the oldest form of an internet.) The charged particles descending through our atmosphere in 1859 got into the telegraph system, and some telegraph operators reported sparks coming out of their instruments, while others saw sparks flying from the telegraph wires overhead. The episode is now called the Carrington event in honor of the person who first connected it with activity on the Sun.

Today, we are far more dependent on electricity than we were in 1859. We had a taste of what can happen in our modern world when a milder geomagnetic storm hit the Earth in March of 1989. Auroras were visible in both Canada and the United States. Electricity and lights went out in Montreal and other parts of Quebec, leaving Montreal's underground system of pedestrian walkways dark until backup lights could come on. The radar at the airport in Dorval stopped working for a while. Elsewhere, people reported their electric garage doors opening and closing randomly.

It just so happened that the Space Shuttle Discovery went into space on March 13 that year, just as the storm was arriving, and the astronauts reported some strange readings on their instruments. A number of satellites already in space reported damage to their electronic instruments because of the surge of electrically charged particles that hit them.

Three-Dimensional Learning Exposed

Many of the experiences in this book emphasize three-dimensional learning when coupled. The best example in this chapter is the combination of Experience 3.4, "Discover the Sunspot Cycle," and Experience 3.7, "What Else Cycles Like the Sun?"

These two experiences use Earth- and space-based data that students analyze to discover patterns in the number of sunspots observed over many years. This information is then compared with data characterizing other phenomena. Students find that some of them are related to the level of sunspot activity (e.g., geomagnetic activity) and some are not (e.g., stock market activity).

We encourage you to find time during these experiences to point out to students how they are engaged in the science practices and crosscutting concepts, especially how what they are doing resembles the practices of scientists—analyzing and interpreting data that have been carefully gathered.

In addition to three-dimensional learning, the experiences in this chapter also make extensive use of mathematics concepts identified in the *Common Core State Standards*, including basic algebra, proportional reasoning, and graphing skills. Reading, writing, and speaking skills are required in many of the experiences, as students use informational text and the results of observations they have made to produce and deliver presentations arguing for their perspectives or ideas.

EXPERIENCE 3.7

What Else Cycles Like the Sun?

Overall Concept

Sunspots are just one manifestation of solar activity. When the number of sunspots increases, so do other phenomena on the Sun and in the solar system. This experience helps students graph one phenomenon on Earth that correlates with the 11-year sunspot cycle and one that doesn't.

Students first graph the rise and fall of the aa index of geomagnetism, which is a measure of the intensity of magnetic activity on Earth. Such activity is known to be connected to the level of solar activity and to the appearance of auroras (Figure 3.18).

Some people have suggested over the years that there may be other characteristics of the natural or human world that are connected to the sunspot (or solar activity) cycle. These suggestions include the ups and downs of the stock market, the quality of good

FIGURE 3.18

Auroras are known to be connected to solar activity.

The aurora from the island of Kvloya in Norway January 23, 2011. When you are at high latitudes, the auroral display seems to come from above you.

MATERIALS

For the groups (each sheet is given to half of the groups):

- "Geomagnetic Activity Data Table" (pp. 224–226)

- "S&P 500 Data Table" (pp. 227–229)

- "Graphing Chart B" (p. 230)

- "Graphing Chart C" (p. 231)

One per student:

- Astronomy lab notebook

French wine, crime statistics, and so on (Figure 3.19). To test one of these suggestions, we provide the S&P 500 stock index (a popular measure of how U.S. stocks are doing) for each quarter and students are asked to make a graph of its rise and fall for about the same period as they used for sunspots in Experience 3.4, "Discover the Sunspot Cycle." We should mention that, so far, none of these measures of human activity have shown any correlation with the sunspot cycle when the data are examined carefully.

Objectives

Students will

1. discuss some of the other measurable consequences of increased or decreased solar activity besides the number of sunspots,

2. make a graph of how some of these (the aa index of geomagnetism and the S&P 500) change

3. determine whether the changes in the aa index do or do not correlate with the sunspot cycle, and

4. determine if the rise and fall of the stock market correlates with the sunspot cycle.

Advance Preparation

1. If possible, display the graphs the class made of the sunspot cycle in Experience 3.4. If there is enough wall space, put it relatively high, so students can put the graphs they make in this experience below it.

2. Make copies of the data tables and graph forms for each of the two quantities students will be analyzing. Half the groups will do one and half the other.

FIGURE 3.19

Many people believe that stock prices are affected by solar activity, but they aren't.

Ringing the bell at the opening of the New York Stock Exchange, September 18, 2012

3.7
EXPLAIN

Procedure

1. Tell the students that the cycle of solar activity which they discovered in Experience 3.4 is one of the key rhythms of our solar system. So, naturally, we are interested in exploring if other things in our environment "keep the same beat"—vary the same way as the sunspots and solar activity. The students will be investigating two possible examples in this experience. As you did in Experience 3.4, you will want to assign a different period of years to each of the groups graphing the same quantity.

2. Some students will examine a measure of the intensity of the Earth's magnetic activity, which determines how many auroras (northern and southern lights) are visible and how far south of the North Pole or north of the South Pole they can be seen. If this has not come up before during one of the previous experiences, explain the connection between an increased number of charged particles coming from the Sun and increased aurora activity at high northern and southern latitudes. You can find some beautiful images of auroras from the NASA Aurora Image Archive to show students at *www.nasa.gov/content/goddard/ purple-and-green-aurora-in-alaska*.

    Tell students that scientists who try to understand the connections between Earth and space (geophysicists) have defined several geomagnetic *indexes*, numbers that measure how strong the magnetic storm activity is that our planet is experiencing and thus how good the chances are of seeing auroras at different latitudes. A number of stations are set up around the world to measure magnetic conditions related to Earth and space, and they all report in so that a worldwide measure of these indexes can be put together. Explain that some of the groups will be getting a record of the aa index for a certain number of years (the number will depend on how many years you want each group to graph) and will be asked to plot it in the same way they plotted the sunspot number in Experience 3.4. Give these groups copies of some or all of the "Geomagnetic Data Table" and "Graphing Chart B."

3. The other half of the students will look at something more speculative. In trying to find explanations for the ups and downs of the stock market, some commentators have suggested that

stocks rise and fall with the same cycle as solar activity. So, these groups will get numbers from a popular measure of the stock market, the S&P (Standard and Poor's) 500, which measures the performance of 500 large companies, many of which are at the heart of our economy ("S&P 500 Data Table", pp. 227–229). These groups will need to use some kind of graph like "Graphing Chart C" because stocks generally tend to rise over time, even while the market goes up and down. So over the years, the S&P 500 grew in value—meaning the *x*-axis of their graph has to have a lot of room for growth as well as for ups and downs. Also note that the record for the S&P 500 in the data table starts in 1950, not 1948 like the sunspot counts.

4. When everyone is done, groups that were graphing different time periods for the same measure (either the aa index or the S&P 500) should combine their graphs so that they cover more time. Then they should compare their graphs to the one done for the solar cycle in Experience 3.4. Groups can take turns pasting their results under the solar activity graph so that the years line up. Ask them to look at the trends on each of the new graphs. They should try to see if each new graph is going up and down in the same years as solar activity does and if not, how it differs. *Students should conclude that the aa index matches the sunspot data and the S&P 500 shows no connection to the sunspot data.*

Teacher note: As they did in Experience 3.4, students will have to resist the urge to simply draw and look at the straight lines that connect the dots on the graph. Instead, they should look for the larger-scale up and down trends on the graph and see if they follow the larger-scale trends for the sunspot counts. This is actually a good experience for students who have not worked with real-world data before but only with idealized sets of numbers in textbooks. In the real world, scientists often find that data are "messy" (could have lots of small-scale ups and downs in them). In some cases, it's only by overlooking the small-scale variations and focusing on large-scale trends that scientific discoveries about long-term changes in our environment or other systems can be made.

5. Encourage students in the class discussion to think and talk about why the phenomena they looked at behave the way they do. Then ask them to speculate what else might be correlated with the 11-year cycle of sunspot activity.

   Teacher note: This is a good opportunity to remind students that science is always a "progress report" and not a giant warehouse of "final answers." For example, this field of examining what connects to the sunspot data is an ongoing area of investigation and debate among scientists. Also, this is an area in which many people outside of science like to make suggestions or press moneymaking ideas. Therefore, if students do further research on their own, they should know that not everything they might read on random websites is necessarily backed up by scientific data. That gives you an opportunity to discuss criteria that scientists use to decide which ideas are worth further investigation and which are not.

   A few other things that keep the same or opposite cycle as sunspots include the following:

   a. The number of flares and coronal mass ejections (CMEs): These are signs of higher solar activity, just as the increase in the number of sunspots is; they increase when the number of spots increases.

   b. Ultraviolet (UV) rays from the Sun: The Sun produces more UV light when there are more active regions, which produces more ozone high in our atmosphere. Ozone blocks UV from reaching the ground, so the amount of UV we measure on the Earth's surface is less when the Sun is more active. (For more on UV, see Experiences 3.8 and 3.9.)

   c. Satellites in orbit show damage or their orbits decay: Satellites are more vulnerable when there is more solar activity, and the greater flow of particles from the Sun leads to more geomagnetic activity.

   d. The arrival at Earth of *cosmic rays* coming from our Milky Way galaxy: These cosmic ray particles are in a "competition" with the particles streaming from the Sun. When solar activity is high, the flow of energetic particles

from the Sun is strong and they keep the cosmic rays out. When solar activity is low and few particles come from the Sun, the cosmic rays have a much greater chance of reaching the Earth.

Teacher note: The term cosmic ray is misleading, but we are stuck with it because it has about a hundred years of tradition on its side. They are not rays at all but particles—atoms or pieces of atoms that come toward Earth from the Milky Way galaxy at large. Many of them come from massive stars that exploded at the end of their lives and sent their atoms at high speed and with high energy into deep space.

Geomagnetic Activity Data Table

This table shows the average aa index value for each quarter from 1950 to 2013.

Year	Month 3	Month 6	Month 9	Month 12
1950	21.0	21.4	26.2	28.7
1951	26.8	26.7	33.0	28.0
1952	34.2	31.6	22.7	22.8
1953	23.6	20.8	25.8	18.3
1954	21.2	14.0	18.2	15.4
1955	20.3	17.5	15.1	17.2
1956	26.4	28.0	20.9	23.2
1957	30.6	25.2	33.0	28.3
1958	34.8	27.4	29.1	22.3
1959	29.9	23.8	36.5	30.3
1960	25.4	36.8	27.2	41.9
1961	22.5	21.3	25.0	20.5
1962	15.9	17.9	25.6	26.3
1963	16.4	19.6	27.8	20.9
1964	20.3	18.0	16.5	13.6
1965	14.0	12.9	16.3	12.8
1966	15.8	13.0	22.1	18.2
1967	17.4	22.3	18.8	20.3
1968	23.6	22.8	19.9	23.7
1969	23.5	21.7	18.0	16.5
1970	17.7	19.3	22.9	19.5
1971	21.8	20.6	17.8	19.8
1972	20.5	18.6	22.8	20.3
1973	31.8	30.9	21.3	22.9
1974	28.5	30.3	31.9	30.4

Page 2

Geomagnetic Activity Data Table (*continued*)

Year	Month 3	Month 6	Month 9	Month 12
1975	30.1	22.5	18.8	23.4
1976	28.3	22.1	19.8	18.5
1977	19.8	19.6	22.9	18.3
1978	25.5	30.2	24.1	22.4
1979	25.9	24.2	21.9	17.7
1980	16.6	18.2	15.9	23.4
1981	22.2	26.2	24.1	26.1
1982	33.4	29.9	40.0	32.3
1983	33.8	31.1	23.5	29.7
1984	27.1	28.1	28.6	31.4
1985	23.1	21.9	22.4	22.6
1986	28.1	16.4	21.3	18.5
1987	16.4	13.6	24.5	21.2
1988	22.9	21.3	20.7	23.6
1989	39.5	27.1	22.7	31.9
1990	33.0	29.0	25.0	19.2
1991	24.9	35.9	35.6	40.6
1992	32.7	24.6	25.9	26.0
1993	32.1	24.4	20.4	25.0
1994	35.9	35.9	18.9	27.0
1995	24.0	24.7	18.0	21.1
1996	20.6	15.2	19.9	18.6
1997	18.2	15.7	14.7	15.8
1998	18.1	21.6	22.5	21.6
1999	21.9	16.6	24.8	25.5
2000	23.6	25.0	28.5	24.4

Page 3

Geomagnetic Activity Data Table (*continued*)

Year	Month 3	Month 6	Month 9	Month 12
2001	21.0	23.0	20.4	25.1
2002	19.0	20.0	21.3	30.6
2003	30.2	42.6	33.2	42.4
2004	29.1	17.5	20.6	25.1
2005	27.3	22.3	26.6	16.6
2006	14.3	15.9	15.1	19.4
2007	16.6	15.2	14.6	13.5
2008	20.3	15.1	10.6	10.9
2009	10.4	8.4	8.6	7.3
2010	9.2	15.7	12.3	11.7
2011	14.0	16.4	17.1	11.7
2012	19.7	17.1	17.5	13.4
2013	15.0	16.7	14.6	13.0

Source: David Hathaway, NASA

NATIONAL SCIENCE TEACHERS ASSOCIATION

S&P 500 Data Table

Year	Month 3	Month 6	Month 9	Month 12
1950	17.4	18.7	19.1	19.8
1951	21.6	21.6	23.5	23.4
1952	23.8	24.4	24.8	26.0
1953	26.0	26.8	26.9	24.8
1954	26.6	29.0	31.5	35.0
1955	36.5	39.8	44.3	45.3
1956	47.5	46.3	46.8	46.4
1957	44.0	47.6	44.0	40.3
1958	42.1	44.8	49.0	53.5
1959	56.1	57.5	57.0	59.1
1960	55.0	57.3	54.8	56.8
1961	64.1	65.6	67.2	71.7
1962	70.3	55.6	58.0	62.6
1963	65.7	70.1	72.9	74.2
1964	78.8	80.2	83.4	84.0
1965	86.8	85.0	89.4	91.7
1966	88.9	86.1	77.8	81.3
1967	89.4	91.4	95.8	95.3
1968	89.1	100.5	101.3	106.5
1969	99.3	99.1	94.5	91.1
1970	88.7	75.6	82.6	90.1
1971	99.6	99.7	99.4	99.1
1972	107.7	108.0	109.4	117.5
1973	112.4	104.8	105.6	94.8
1974	97.4	89.8	68.1	67.1
1975	83.8	92.4	84.7	88.7

S&P 500 Data Table (*continued*)

Year	Month 3	Month 6	Month 9	Month 12
1976	101.1	101.8	105.5	104.7
1977	100.6	99.3	96.2	93.8
1978	88.8	97.7	103.9	96.1
1979	100.1	101.7	108.6	107.8
1980	104.7	114.6	126.5	133.5
1981	133.2	132.3	118.3	123.8
1982	110.8	109.7	122.4	139.4
1983	151.9	166.4	167.2	164.4
1984	157.4	153.4	166.1	164.5
1985	179.4	188.9	184.1	207.3
1986	232.3	245.3	238.3	248.6
1987	292.5	301.4	318.7	241.0
1988	265.7	270.7	268.0	276.5
1989	292.7	323.7	347.3	348.6
1990	338.5	369.4	315.4	328.8
1991	372.3	378.3	387.2	388.5
1992	407.4	408.3	418.5	435.6
1993	450.2	448.1	459.2	466.0
1994	463.8	454.8	467.0	455.2
1995	493.2	539.4	578.8	614.6
1996	647.1	668.5	674.9	743.3
1997	792.2	876.3	937.0	962.4
1998	1,077	1,108	1,021	1,190
1999	1,282	1,323	1,318	1,429
2000	1,442	1,462	1,468	1,331
2001	1,186	1,231	1,045	1,145

Page 3

S&P 500 Data Table (*continued*)

Year	Month 3	Month 6	Month 9	Month 12
2002	1,154	1,014	867.8	899.2
2003	846.6	988.0	1,019	1,081
2004	1,124	1,133	1,118	1,199
2005	1,195	1,202	1,226	1,262
2006	1,294	1,253	1,318	1,416
2007	1,407	1,514	1,497	1,479
2008	1,317	1,341	1,217	877.6
2009	757.1	926.1	1,045	1,110
2010	1,151	1,083	1,122	1,242
2011	1,304	1,287	1,174	1,243
2012	1,389	1,323	1,443	1,422
2013	1,551	1,619	1,687	1,808
2014	1,864	1,947	1,993	N/A

Source: Political Calculations. The S&P 500 at your fingertips. *http://politicalcalculations.blogspot. com/2006/12/sp-500-at-your-fingertips.html#.ViVM7CtmKYx.*

Graphing Chart B

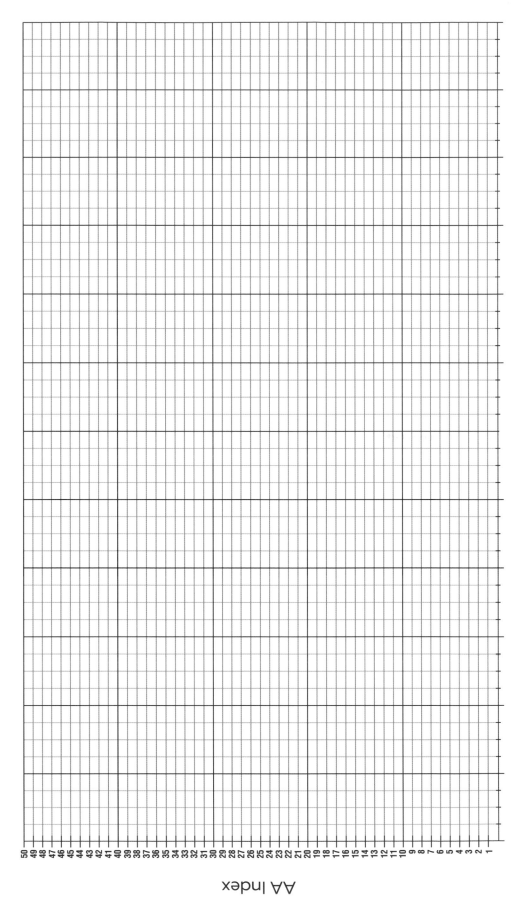

YEARS

AA Index

S&P Index

2000 1950 1900 1850 1800 1750 1700 1650 1600 1550 1500 1450 1400 1350 1300 1250 1200 1150 1100 1050 1000 950 900 850 800 750 700 650 600 550 500 450 400 350 300 250 200 150 100 50

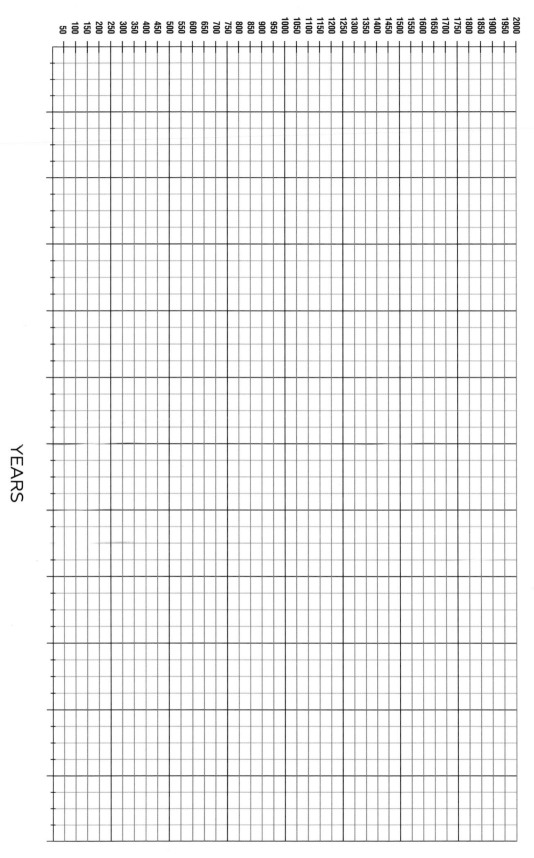

YEARS

Graphing Chart C

EXPERIENCE 3.8

The Multicolored Sun

AN INTRODUCTION TO ELECTROMAGNETIC RADIATION

Overall Concept

Students begin the experience with a take-home challenge of understanding how a TV remote communicates with a TV. This leads to an introduction of the idea of a spectrum of "invisible light," or electromagnetic radiation. In class, students brainstorm about other technology in their lives that involves invisible light. After learning more about the electromagnetic spectrum and the bands into which scientists divide it, they examine and discuss images of the Sun taken the same day but in several different bands of the electromagnetic spectrum (e.g., x-rays, ultraviolet [UV], and visible light).

Objectives

Students will

1. understand that technologies in their everyday lives involve invisible beams of radiation;

2. understand that visible light is only one example of a much larger family of *electromagnetic waves*;

3. understand that the Sun can also be observed in some of these electromagnetic waves, especially when scientists use instruments in space; and

4. start to understand why images of the Sun taken in various electromagnetic waves look different.

MATERIALS

One per group:

*   "Sheet of Solar Images for Experience 3.8," with photos taken on the same day at different wavelengths (p. 242)

One per student:

*   "TV Remote Take-Home Assignment" (pp. 240–241)

*   Astronomy lab notebook

Advance Preparation

1. The take-home activity with TV remotes should be assigned a few days before the class will do the rest of the experience.

2. If you are going to use the images we suggest under the Procedure to teach about the other bands of the electromagnetic spectrum besides visible light, you may want to get the images into a PowerPoint presentation or use some other program from which you can project them.

Procedure

1. A few days in advance, give each student a copy of the "TV Remote Take-Home Assignment" handout and encourage them to involve their families in doing the activities described.

2. Have each student attach the completed handout to a new page in his or her astronomy lab notebook. Then have students break into small groups and discuss and compare the results of their experiments with the TV remote. When the group is in agreement, have them write up their conclusions.

3. Have the groups report out to the whole class and allow for different voices and ideas to be heard. Then you can explain that the remote beam consists of *infrared* waves (waves like visible light but longer in wavelength than the longest red color our eyes can see). These invisible waves travel between the remote and the TV. They are not able to go through walls, so they are limited to the room in which they are being used. The infrared waves can be reflected by a mirror, just as visible light can be reflected, providing further evidence that they are like light.

4. (*Optional*) Show students the following pair of images, which present a scientist at the Jet Propulsion Laboratory with his arm wrapped in a plastic garbage bag (Figure 3.20, p. 234). (You can mention to your students that generally scientists spend very little of their time with their arms inside garbage bags; this was for the sake of this special demonstration.) If you want to project it, the image can be found at *https://dlnmh9ip6v2uc.cloudfront. net/assets/8/c/5/2/1/511917bbce395fef32000000.jpg*.

FIGURE 3.20

A pair of photos showing the same scene in visible and infrared light

The first picture is a visible light image, while the second shows infrared or heat rays, which tell us how warm things are. Have the students get back in their groups and discuss the differences between the two images and what we can learn from the infrared picture. Have groups report their ideas to the whole class. Also make a list of questions they have after this section.

5. (*Optional*) Ask students to return to their small groups and think about whether they have ever had a photo taken of their bodies that didn't involve visible light. Most likely, someone will think of an x-ray of their bones or their teeth.

Teacher note: In some cases, students who have had an illness or been to the hospital with someone who has been ill or pregnant might think of ultrasound or MRI exams. But as the groups report out, focus on x-rays, which are likely to be more familiar. (For your reference, MRI scans involve magnetic fields and radio waves. Ultrasound uses sound—air pressure waves— at frequencies so high humans cannot hear them.)

Ask students to discuss what x-rays of their bodies show that regular photographs can't. They should eventually come up with

the idea that x-rays don't show your skin or flesh, just the bones inside. You could show them the photograph in Figure 3.21. If you want to project it, you can find it at *http://upload.wikimedia.org/ wikipedia/commons/f/ff/X-ray_1896_nouvelle_iconographie_ de_salpetriere.jpg*. (The 1896 pair of images shows a patient with one finger shorter than the rest, seen with light and with x-rays. It is one of the earliest x-ray images we have.)

6. Ask the students to get back into their groups and brainstorm about other ways they have interacted with invisible waves during their lives. If students are really stuck, take out a cell phone and just wordlessly look at it and then put it away. This usually gets them thinking about ways we interact using invisible waves. After some discussion, ask each group to report out and make a list on a board or poster for the entire class to refer to.

FIGURE 3.21

A pair of images showing a visible light photo and an x-ray image

 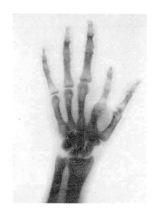

Teacher note: Among the many possible answers are listening to the radio (at home or in their cars), watching TV, talking or playing games on a cell phone, using a wireless tablet or laptop computer, heating up a meal or drink in a microwave oven, having an outside door in a bank or office open automatically without touching it, and getting a tan. If they didn't do the optional part in step 5 above, someone may also think of getting an x-ray at a doctor or dentist's office.

7. Select your favorite way for students to learn more about the entire electromagnetic spectrum. You might have a class discussion, assign homework, or have different student groups research each band of the spectrum (e.g., radio, microwaves, infrared, visible light, UV, x-rays, gamma rays) and report back to the whole class.

3.8
ELABORATE

Teacher note: One useful resource is the NASA Mission: Science page "Electromagnetic Spectrum Tour," found at *http://missionscience.nasa.gov/ems/index.html*. For more on how infrared and UV (the first invisible electromagnetic waves to be known) were discovered, see the "A Little Infrared and UV History" box at the beginning of the chapter (pp. 169–170).

8. One key concept to get across is that objects at different temperatures emit different kinds and amounts of electromagnetic waves. At classroom temperatures, your students and their notebooks primarily give off infrared waves. At the temperatures of thousands of degrees on the surface of the Sun, the vibrating atoms primarily give off visible light. In the heated outer atmosphere of the Sun (called the corona), even hotter atoms primarily give off UV waves and x-rays.

9. Now that students understand that images of the Sun taken with different kinds of light can show different things about the Sun, give out or project on a big screen one or more images of the Sun in other bands of the spectrum. These can either be current (using the websites below) or you can use the images on the "Sheet of Solar Images for Experience 3.8." Ask the students (in their groups) to brainstorm what these images not taken with visible light could be telling us about the Sun. They should also make a list of the questions they have.

   Teacher note: Current images of the Sun can be found every day at a number of websites:

   • Solar Monitor: *www.solarmonitor.org*

   • Solar Data Analysis Center: *http://umbra.nascom.nasa.gov/newsite/images.html*

   • *Helioviewer.org: http://helioviewer.org* (click the help button at the bottom for instructions)

   Students will generally have no reference from their everyday experience to help them evaluate these images, except the example with the x-rays above. You will need to be prepared for some novel ideas and questions related to the images.

10. After the students have brainstormed for a while in small groups, bring everyone together and have the groups report their ideas to

the class. This is not the time to get every fact right but simply to list the thoughts students have on the whiteboard or poster sheet. You can also list any questions they have for future discussion. Students will return to examining the Sun at many wavelengths in some of the later experiences (Figure 3.22).

11. After the students have had a chance to share their ideas, discuss a bit more with them about how solar scientists learn about

FIGURE 3.22

The Sun at the many wavelengths that the Solar Dynamics Observatory is able to observe

Note that the *A* with the little circle over it is the symbol for an Angstrom unit, a small unit of length used for measuring the characteristics of atoms and electromagnetic radiation (light). One Angstrom unit is one hundred millionth of a centimeter.

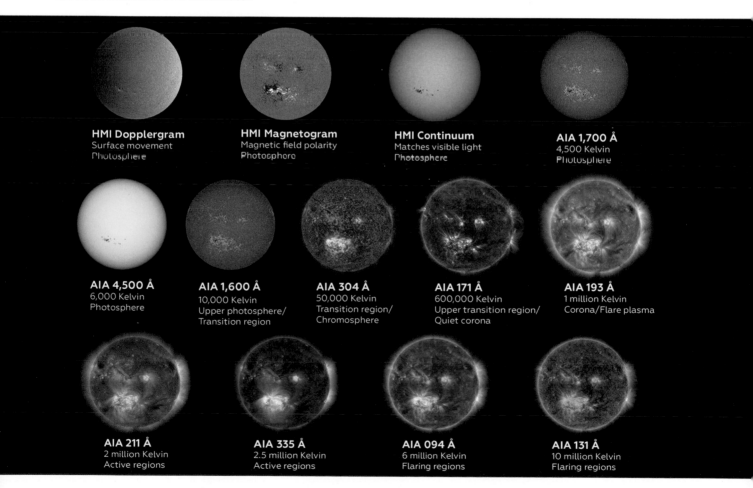

the Sun at different wavelengths. (See the Content Background section [pp. 170–171] and the Teacher Note below for more information and resources.) The following are key ideas:

a. Visible light images show us the photosphere of the Sun and allow us to count sunspots and sunspot groups (see top left photo on "Sheet of Solar Images for Experience 3.8").

b. When we take images of the Sun in UV light, we see the hotter regions above the photosphere where flares and other examples of solar activity are more easily visible. For example, the Solar Dynamics Observatory spacecraft can take images in what scientists call the *extreme UV* region of the spectrum. These UV waves come from areas in the atmosphere of the Sun that are at temperatures of millions of degrees, temperatures reached by the Sun's corona (or outer atmosphere) but not by the gas at the Sun's surface (see bottom right photo on "Sheet of Solar Images for Experience 3.8").

c. X-rays show even hotter regions in the most energetic part of the Sun's corona (see top right photo on "Sheet of Solar Images for Experience 3.8").

d. Astronomers want to monitor the hottest regions of the Sun because they are the source of coronal mass ejections (see Experience 3.6, "Space Weather: Storms From the Sun," for more on these) that may threaten the Earth.

Teacher note: After students have done their initial think-pair-share activity, you could show them or send them to some useful websites to learn more about the Sun at different wavelengths:

a. The Cool Cosmos site from IPAC (the Infrared Processing and Analysis Center, run by the California Institute of Technology's Jet Propulsion Laboratory) has a multiwavelengths Sun webpage that is full of good information at *http://coolcosmos.ipac.caltech. edu/cosmic_classroom/multiwavelength_astronomy/ multiwavelength_museum/sun.html*. (Students can see

3.8
ELABORATE

pictures of the Sun, starting with x-ray images all the way down to radio images. If students click on any of the image boxes, they will be taken to a background page which explains more clearly just what can be seen on each image.)

b. The website of the Solar Dynamics Observatory spacecraft, where current images of the Sun in different wavelength can be found is at *http://sdo.gsfc.nasa.gov/data*. (Note that students can click on the little "i" symbol in the upper right corner of each picture box, which will pop up a box of text with more information about what kinds of waves and what kinds of temperatures each image is showing.)

c. NASA's "Sun Primer: Why NASA Scientists Observe the Sun in Different Wavelengths" (a bit more technical) is at *www.nasa.gov/mission_pages/sunearth/news/light-wavelengths.html#.VMl5eGjF98E*. (Be sure to tell students to scroll down to the bottom of the page where the Sun images at different wavelengths are. If you click on any of the small Sun images, you get more information.)

3.8

TV Remote Take-Home Assignment

(If possible, involve your whole family in doing this project, and have fun with it.)

━━━━━━━━━━━━━━━ ☀ ━━━━━━━━━━━━━━━

When you will not be disturbing someone in the middle of a favorite TV show, ask your parents if you can do some experiments with the remote control that turns on your TV set. First, make sure the remote is working (has batteries that still have some power).

1. Turn the TV off and go to the farthest distance in the room from the TV. Point the remote at the TV. Does it turn on the TV? _____ If not, how close do you have to be for the remote to work?_____

2. Turn off all the lights in the room (this is best done at night) and then point the remote at the TV to turn it on. Can you see any beam coming from the remote when the room is dark? _____ (In general, does the remote work any differently when the room is full of light and when it is dark? _____)

3. Now try putting some things in front of the remote, between the remote and the TV, and see if the beam makes it through to turn on the TV. It's nice to have someone in the family helping you with this part. Does the remote beam make it through the materials in the chart below? (You can select some other material to fill in the last row in the box.)

Material in front of remote	Does the beam go through?
A sheet of notebook paper	
An empty transparent glass	
A glass full of water	
A glass full of milk	
Aluminum foil	
A person's hand	

4. Now get a hand mirror from your parents and someone to help you by holding the mirror. See if you can angle the mirror to turn on the TV not by pointing the remote directly at the TV but by pointing it at the mirror and then having the mirror reflect the beam to the TV. (This is tricky, because the mirror angle has to be just right, but it's fun to show off when it works.)

Page 2

TV Remote Take-Home Assignment

---

5. What conclusions can you draw about the beam that comes from your remote to turn on and control the TV?

_____

_____

_____

_____

_____

_____

_____

_____

_____

_____

_____

_____

_____

_____

_____

_____

_____

_____

_____

_____

_____

_____

Sheet of Solar Images for Experience 3.8

All of these images were taken on April 18, 2015

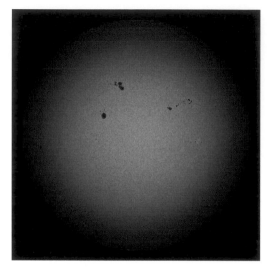

Visible light from the
Solar Dynamics Observatory

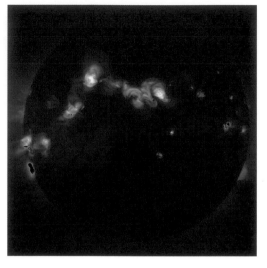

X-rays from Hinode spacecraft

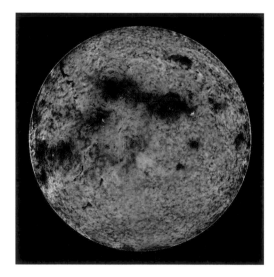

Infrared from
National Solar Observatory

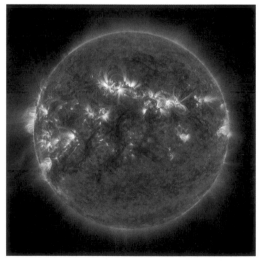

Extreme UV from the
Solar Dynamics Observatory

(Note that all these images are artificially colored. Don't pay attention to the colors, but
look at the structure—the details of what is bright and dark and where things are.)

EXPERIENCE 3.9

Student Detectives and the Ultraviolet Sun

Overall Concept

Students continue their exploration of the electromagnetic spectrum by examining the invisible ultraviolet (UV) radiation coming from the Sun using UV-sensitive beads. Student groups are given the challenge of seeing how much different substances can block UV waves. They also discover that scattered UV rays are found even in the shade.

Objectives

Students will

1. understand that the Sun gives off invisible UV waves (most of which are absorbed by the ozone layer, but some of which can still be detected on the Earth's surface);

2. appreciate that different materials block the Sun's UV rays to different degrees; and

3. understand that UV waves can be scattered by air molecules into shady areas where there is no direct sunlight.

Advance Preparation

1. This activity is most effective if you let students experiment with a wide range of materials and situations; thus, we recommend beginning to gather the needed materials a week or two in advance. Students and parents may be able to help collect some of the materials.

MATERIALS

For each group:

*   UV-detecting beads (at least a dozen; some teachers like to reward students with beads to take home, in which case you want at least six per student)

    Teacher note: Beads can be ordered from Educational Innovations at *www.teachersource.com/product/ ultraviolet-detecting-beads/light- color.* (The beads are off-white in the dark, but they turn different colors when exposed to UV rays. Some teachers prefer beads of just one color, especially deeper colors such as purple, while others like an assortment.)

*   Storage container (at least one) that blocks UV light

    Teacher note: Brownish-orange cylindrical pill bottles seem to work well. (You might ask students to collect these a week or two in advance by asking parents if they have any that are empty and have been washed.)

(continued on p. 244)

2. Beads should be ordered far enough ahead to arrive when you need them.

3. You need to determine whether there is a place in your classroom where direct sunlight comes in or if you will need to do the experience outside.

Procedure

1. Discuss or review with the class that the Sun gives off not just visible light but also radiation that is invisible to human eyes. In this experience, the class will focus on UV rays, which have more energetic waves than visible light. Many of the more energetic things happening on the Sun can be observed in UV light.

2. Discuss with the class the dangers that UV radiation poses to humans: It can burn lighter-colored skin, damage eyes, and in larger quantities, cause skin damage and cancer. Luckily, there is a layer high in the Earth's atmosphere that keeps out most of the UV coming from the Sun. It is called the *ozone layer* and it is roughly 20–30 km (12–18 mi.) above the Earth's surface. When astronomers want to observe UV rays from the Sun to understand what is happening in the Sun's outer layers, they launch specialized UV telescopes into orbit, so they are above the

MATERIALS (*continued*)

- A sunny location near an open window or outside

- 1 regular flashlight

- 1 UV bulb or light (sometimes called a black light)

- Sunscreen lotion; if you have the resources, you can provide sunscreens with three different SPF (Sun protection factor) numbers for each group

- Resealable plastic bags (several)

- 1 marker that can write on plastic bags

- 1 cup or beaker of water for each group

- 1 sheet of white construction paper

- 1 sheet of black construction paper

- 1 box of tissues

- 1 pair of sunglasses (students could bring these from home)

One per student:

- "Ultraviolet Detectives Record Sheet" (pp. 248–249)

- Astronomy lab notebook

- (*Optional*) Pipe cleaners, ribbons, shoelaces, or the like to string beads on, if you want students to take beads home and not lose them

Earth's ozone layer. Even though the ozone layer absorbs much of the UV radiation from the Sun, some of it gets through and reaches us on our planet's surface. This is why doctors say that you should put sunscreen lotion on if you spend significant time in direct sunlight.

3. Since our eyes can't see UV, we need a "detector" of some kind that will reveal the presence of UV. The beads we are going to use as UV detectors contain a special pigment that turns color when UV hits them but not when regular light does. The more UV hits them, the darker they turn. The beads return to being colorless once they have been away from UV for a while.

   Teacher note: For the technical details about how the beads work, see *www.arborsci.com/Data_Sheets/P3-6500_DS.pdf*. Cooling the beads in ice water, for example, returns them to clear more quickly.

4. Explain that the transparent brownish bottles that many medicines are kept in let visible light through, so you can see how many pills are left, but don't let UV through, since UV can harm some medicines. Bring this up if you are using these bottles to keep the beads from being exposed to UV before you are ready to use them.

5. Next demonstrate how the UV-sensitive beads work. Take one out of a medicine bottle and, while keeping it out of sunlight, shine a flashlight on it. Note that light alone doesn't make the bead change. Now expose it to UV by putting it into sunshine near an open window (window glass often keeps out UV) or next to a UV light (black light). Everyone should see the bead turn from clear to colored.

6. Now challenge the class by officially converting them into squads of UV detectives. Their mission is to figure out what can keep UV away from where it's not wanted. They should test a wide range of materials to see which ones make their beads change color and which ones don't. Have students work in groups, but give each student his or her own "Ultraviolet Detectives Record Sheet."

7. Before students are dismissed to work in their groups, point out a few of the substances you have available for them to test

(although they are by no means limited to those substances). Because sunscreen lotion can be messy and may get on clothes and school materials, it's probably best to develop a protocol for testing the one or more lotions you have available for them. Many classes have found the easiest way to deal with sunscreen lotions is to spread the lotion inside a small resealable bag and either put the whole bag on top of the bead or insert the bead into the lotion inside the bag. If you have lotion with different SPF values, it's important to label each bag with the SPF (using a marker). In some classes, the teacher prepares the lotion baggies and then has each group take turns with them, but it's nice if each group can have its own lotion bag to experiment with, especially if they are going outside to do the testing.

8. Explain where the students can go (the classroom, a hallway with an open window, outside in the school yard, etc.) and how much time they have for their experiments. Encourage them to bring pencils with them to record their observations on the handout while it's fresh in their minds.

9. If you want, you can tell students that they will be allowed to make a small bracelet of beads later and take it home for further experiments. That way, they will not be discouraged if the class time doesn't allow them to test as many substances as they want.

10. One of the interesting experiments the students can be encouraged to try is to go outside and, in direct sunlight, watch a bead get exposed to UV and change color. Then they should take a fresh bead and, keeping their hands around it, go into the shadow of a building or a tree before exposing a bead while remaining completely in the shadow. They should observe that the bead still changes color. Ask them why this might happen. (*Molecules of air in our atmosphere scatter the UV radiation in all directions, so that some UV is everywhere sunlight has scattered.*)

11. You can, if there is time, encourage your students to investigate sunblock lotion more carefully. Is it just the presence of the white opaque lotion that blocks the UV, or is it some specific chemical in the lotion? Cream cheese is a white substance that you can smear inside a resealable bag in a way similar way. It will generally not block UV even though regular light doesn't get through it.

3.9
ELABORATE

12. After the groups have finished their investigations (this could be the next school day or even the next week), have them report out to the whole class, discussing what their detective work has revealed. (It's best to take turns, so each group reports on one substance and how well it shielded UV before it's the next group's turn.)

13. Students should record their final conclusions in their astronomy lab notebooks.

14. If it fits with your school procedures, encourage the class to set up a library exhibit or bulletin board showing what they found and how it will impact their behavior and their recommendations to others.

Ultraviolet Detectives Record Sheet

※

Your name: _____ Date: _____

Others in your group: _____

Ultraviolet (UV)-sensitive beads are a good way to tell whether substances block the Sun's UV radiation. Please test as many things as possible, first predicting whether the substance will block UV and then actually testing each one. You can try different types of clothing, for example. Remember to keep the bead in the medicine bottle or other covered container that you got them in until you are ready to use them. If you go outside, make sure you hold the bead or beads you plan to use in such a way that they are completely covered by your hand or clothes.

Use these symbols to fill out the chart: C = clear, SC= slightly colored, DC = deeply colored.

Substance being tested	Your prediction for the bead	Your observation of what happened to the bead
Direct sunlight (no window in the way)		
Sunlight through a closed window		
Outside air in the shade		
Sunlight through clouds		
Brownish pill bottle		
Sunscreen lotion, SPF =		
Sunscreen lotion, SPF =		
Sunscreen lotion, SPF =		
Eyeglasses (clear)		
Sunglasses		

Page 2

Ultraviolet Detectives Record Sheet

✳

Substance being tested	Your prediction for the bead	Your observation of what happened to the bead
Baseball cap		
White construction paper		
Black construction paper		
Tissue paper		
A cup filled with water		
Windshield glass from a car		

Fill in any other substances you decided to test in the empty rows.

QUESTIONS

1. If a student is really sensitive to UV rays, what recommendations would you make to that student?

2. Can a student with light skin get a sunburn on a cloudy day? How do you know?

3. If you did the sunblock lotion and cream cheese experiment, explain what you observed.

EXPERIENCE 3.10

Additional Ways of Observing the Sun Safely

Overall Concept

Students learn and practice other methods for safe Sun viewing besides the one explored in Experience 3.3, "Safe Solar Viewing: Project and Record Your Own Images of the Sun." These are particularly useful for times when a partial or total solar eclipse is visible.

Objectives

Students will

1. learn different ways to safely observe the Sun and

2. practice using some of these ways.

Advance Preparation

The first time you do this, you need to order the eclipse glasses a month or two in advance if you decide to do this part of the experience. You need to make sure you get them from a reliable company that has experience making them. The following are good sources:

- Rainbow Symphony: *www.rainbowsymphony.com/ eclipse-glasses.html*

- Thousand Oaks Optical: *www.thousandoaksoptical. com/ecplise.html*

- American Paper Optics: *www.3dglassesonline.com/ our-products/eclipsers*

MATERIALS

Per group:

- 2 pieces of white cardboard (8 ½ × 11 in.; if you can't get enough cardboard, you can also use plain white paper)

- 1 straight pin for making pinholes in cardboard

- 1 or 2 squares of aluminum foil for each group

- A small mirror, such as a hand mirror or makeup mirror for a purse

- Cardboard to cover the mirror

- Scissors

- "A Safe Way to Observe the Sun" instruction sheets (choose which to use with your students): "Pinhole Projection Using Two Sheets," "Pinhole Projection in a Box," "Projection Using a Mirror," and "Using Special Eclipse Glasses" (pp. 253–256)

- (*Optional*) 1 closeable shoe box (or other long narrow box)

One per student:

- (*Optional*) Mylar eclipse glasses

If you are going to do the optional section (making a pinhole projector inside a box), you need to get boxes for each group to work with. The longer the box, the better this works. Shoe boxes are often easiest to get (students can ask parents to see if they have some they can donate to the project). The astronomers at the Exploratorium museum in San Francisco have discovered that taping together two triangular UPS shipping tubes works really well.

Procedure

Safety note: Remind students that it is not safe to look directly at the Sun.

1. Remind students that the Sun is dangerous to look at without a way to protect your eyes. Even when a part of the Sun is eclipsed by the Moon, the rest of the Sun is just as dangerous. This is why it's really important that any class studying the Sun learn Sun-viewing safety. There are safe ways to look at the Sun, either by putting something protective between it and your eyes (a filter, special glasses, etc.) or by projecting an image of the Sun (where the light is strongly reduced and thus safe).

2. If you did Experience 3.3, which used binoculars to project an image of the Sun onto paper, remind the students of that. But since it's not always possible to have a pair of binoculars, explain that you want them to learn other ways of seeing what's happening with the Sun. These won't give as large and clear an image as the binoculars projection (or allow you to see sunspots), but they are still useful, especially when there is a partial eclipse of the Sun (and you want to see the "bite" the Moon takes out of the Sun's disk).

   Teacher note: Eclipses will be discussed in Chapter 4.

3. Distribute the instruction sheet(s) and materials. Tell the students that they will follow the instructions on their sheet to use a certain method to view the Sun. You can circulate among the groups, giving them help as they need it and encouraging them to move on to the next method—if you do more than one—when the time is right.

   Teacher note: The pinhole projection box is the same idea as the pinhole without the box but provides a dark space to see

the Sun's image more easily. For some teachers, finding enough boxes is a problem. Instructions with photographs on projecting an image using a pinhole can be found at the Jet Propulsion Lab site at *www.jpl.nasa.gov/education/index.cfm?page=341*.

4. If you purchase eclipse glasses, it is probably best to wait to distribute the glasses until the students have finished the other parts of the experience. Those glasses provide such an easy viewing experience, they should be saved as a reward for when the students finish their other work.

A Safe Way to Observe the Sun

PINHOLE PROJECTION USING TWO SHEETS

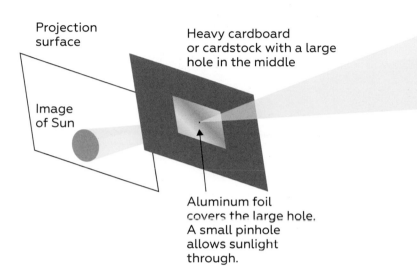

Projection surface

Heavy cardboard or cardstock with a large hole in the middle

Sun

Image of Sun

Aluminum foil covers the large hole. A small pinhole allows sunlight through.

1. Take two sheets of white cardboard or paper and cut a small square out of one of the sheets.

2. Tape a piece of aluminum foil over the hole. Then, make a pinhole at the center of the foil.

3. Stand with your back to the Sun (facing in the opposite direction as the Sun). Hold the sheets up so the pinhole sheet is closer to the Sun. The pinhole acts like a lens, and you will then see a small image of the Sun on the other sheet. (If the Sun is high in the sky, you can put the sheet without a hole on the ground and then hold the pinhole sheet between the ground and the Sun.)

4. The farther apart the two sheets are, the larger the image you will see.

5. This method doesn't show you much detail, but it is useful during a partial eclipse to see the "bite" the Moon takes out of the Sun.

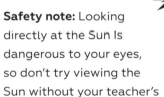

Safety note: Looking directly at the Sun Is dangerous to your eyes, so don't try viewing the Sun without your teacher's guidance.

A Safe Way to Observe the Sun

PINHOLE PROJECTION IN A BOX

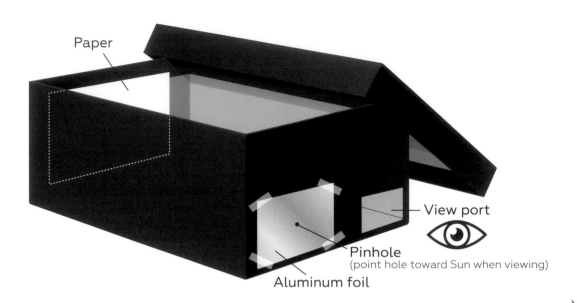

Paper

View port

Pinhole
(point hole toward Sun when viewing)

Aluminum foil

1. Use a shoe box or a long cylindrical or triangular tube that a mailing service might use to mail rolled up posters or prints. The longer the box, the bigger the image will be. To make a longer box, you can tape together two boxes and just remove the dividing wall between them.

2. Now cut a hole in one end of the box and tape a piece of aluminum foil over the hole.

3. With a pin, make a small hole in the middle of the aluminum foil.

4. Tape a piece of white paper to the opposite end of the box. This is where the image will be.

5. Cut a large rectangular hole on one side of the box so you can look at the image projected from the pinhole to the opposite side (see diagram). If the shoe box has no cover, the open side can be your viewing portal.

6. Stand with the Sun behind you. Point the pinhole end of the box toward the Sun until, looking through the opening on the side, you see an image of the Sun on the opposite end from the pinhole. The longer the box is, the larger the image of the Sun will be.

7. Be careful not to look at the Sun! The idea is to have the pinhole pointing at the Sun and your eyes pointing at the image in the opposite direction.

8. This method doesn't show you much detail, but it is useful during a partial eclipse to see the "bite" the Moon takes out of the Sun.

Safety note:
Looking directly at the Sun is dangerous to your eyes, so don't try viewing the Sun without your teacher's guidance.

A Safe Way to Observe the Sun

PROJECTION USING A MIRROR

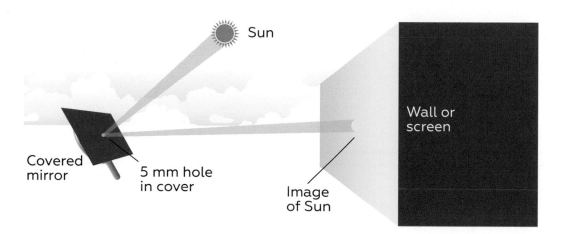

Sun

Wall or screen

Covered mirror

5 mm hole in cover

Image of Sun

1. Get a small mirror (like a hand mirror) and make sure the surface is clean.

2. Cover the mirror with some cardboard in which you made a small hole (no bigger than 5 mm, or ⅕ in.). The rest of the mirror should be completely covered by cardboard or paper.

3. Find a room that has a window in which you can see the Sun (so sunlight hits the window sill). Alternatively, you can be outside and have the Sun on one side of you and a wall that is in shadow on another side.

4. The trick is to hold the mirror so that you catch the sunlight and project it on a dark wall of the room or on a wall outside that is in shadow. This takes some practice, so don't get discouraged.

5. Another student can hold a white piece of paper or cardboard right where the Sun's image is on the wall so it can be seen more clearly.

6. You get the sharpest image when the hole through which the mirror shows is really small and the wall is farther away, but this also produces a fainter image, which can make it harder to see. It is good to experiment with different sized holes and distances to the wall to determine how to get the best results.

7. This method doesn't show you much detail, but it is useful during a partial eclipse to see the "bite" the Moon takes out of the Sun.

8. As a bonus, you can experiment to see whether the shape of the hole makes any difference to the image of the Sun on the wall.

Safety note: Looking directly at the Sun is dangerous to your eyes, so don't try viewing the Sun without your teacher's guidance.

A Safe Way to Observe the Sun

USING SPECIAL ECLIPSE GLASSES

1. Usually made of paper, these glasses remind people of some of the ones you get when you go to a 3D movie, but these have special lenses that protect your eyes. The part of the glasses in front of your eyes is made of a dark material (Mylar, black polymer) that is effective at cutting down the rays of the Sun to a level that does not hurt your eyes.

2. Before using the glasses, check the dark material for scratches, holes, or other damage and discard any glasses that have problems. Also, be sure that you put on the glasses snugly, putting the handles behind both ears. You don't want them falling off when you move or turn your head.

3. Note that if there are really large sunspot groups on the Sun, sometimes you can see them with just your eyes when you are wearing these glasses.

Safety note:
Looking directly at the Sun is dangerous to your eyes, so don't try viewing the Sun without your teacher's guidance.

EXPERIENCE 3.11

Space Weather Report

Overall Concept

Students in small groups are told they will be auditioning for the evening TV news of the future and have to produce a brief "space weather report" segment in the style of today's newscasts. Students are provided with (or find) websites for getting the most current information, but it's up to them to write a report or make a presentation that ordinary people watching TV news can understand. They are encouraged to share their report with their families.

Objectives

Students will

1. demonstrate their understanding of the phenomenon of space weather; and

2. communicate this understanding to others (including those not in the class).

Advance Preparation

1. Plan ahead to give students enough time (in group discussion and for homework) to prepare their segments.

2. For younger students, make a library of resource materials, including images, available to them (see suggestions below and at the end of the chapter).

MATERIALS

For the class

- A setup at the front of the classroom (if only a desk) that simulates a TV studio

- (*Optional*) Video camera (separate or on a smartphone)

- (*Optional*) A computer and LCD projector to show images or project a video (if student groups prepare one)

For each student:

- Students' notes from a previous experience in the classroom about space weather (such as Experience 3.6, "Space Weather: Storms From the Sun")

Procedure

1. Students in small groups review what they have learned about space weather in previous experiences or other class sessions. They also read or do web research about space weather as necessary.

2. Depending on the age and ability of the students, you may want to let them find materials on their own or lead them to the following websites (or others like them). Younger students may need guidance on how to use the information these sites provide:

   a. Space Weather Prediction Center: *www.swpc.noaa.gov*

   b. SpaceWeather.com, which consolidates information from many sources: *www.spaceweather.com*

3. Working in their small groups, students write and produce a three- to five-minute space weather segment for the evening TV news. If the current space weather is very boring, they can choose another date to report on, but their information must be based on real data from a particular day (and they should include the date in their report). If their report includes a particularly large flare or coronal mass ejection, they should feel free to comment on precautionary measures that people or space agencies should take.

4. Students should not assume that their viewers are familiar with the technical terms about space weather that they have learned in class and should try to use everyday words in their reports.

5. Students can use this as an opportunity to be creative and can include posters, PowerPoint slides, or other ways of showing visuals—and even some appropriate humor or music.

6. If video cameras are available, students can make or include a video of some of their data or do a video of the entire segment.

7. After student groups have had enough time in class and at home to work up their reports, each group presents its report (audition segment) live or recorded to the whole class. Students gently "critique" each audition segment, making suggestions for concepts that may need more explaining, terms that the viewers may not know, and so on.

8. Once students feel comfortable with their weather reports, they can share them with families and friends.

EXPERIENCE 3.12

Predict the Next Sunspot Maximum and Minimum

Overall Concept

Using the information from *explore* Experience 3.4, "Discover the Sunspot Cycle," students predict when the next maximum time and the next minimum time for sunspots will be. As an option, they can predict what other phenomena will be at maximum when the Sun is at maximum activity (from *explain* Experience 3.7, "What Else Cycles Like the Sun?"

Objectives

Students will

1. demonstrate their understanding of the 11-year sunspot cycle, and

2. *(optional)* show their understanding of what other phenomena have the same rhythm as the sunspot cycle.

Advance Preparation

1. Keep the class chart from Experience 3.4 up on the wall or return it to the wall when you are ready to do this activity.

2. If you are doing the optional part, also keep or return any charts made in Experience 3.7.

Procedure

1. Ask students to review the notes in their astronomy lab notebooks about the sunspot cycle.

MATERIALS

For the class:

- Graph of the sunspot cycle constructed in Experience 3.4 or "Averaged Sunspot Numbers for the Past Few Decades" (p. 262)

- *(Optional)* Graphs from Experience 3.7

One per group:

- Ruler

One per student:

- Astronomy lab notebooks

2. Have students divide into groups and take turns coming up to the chart they made earlier of the sunspot cycle over the years. Ask them to use their rulers to measure the length of the last full sunspot cycle in centimeters or inches and calculate how many centimeters or inches equal one year on the graph. This allows them to measure the distance between sunspot maximums in centimeters or inches and convert to years.

3. If they did not do this earlier, you can now ask them to return to the graph in groups and check if the length of each sunspot cycle has been exactly the same. Assign a different time period to each group over which to make measurements of the length of the sunspot cycles and then report out to the whole class. With this data in mind, ask students how they might best predict the next solar maximum. If they don't come up with it themselves, lead them to the conclusion that they need to find the average of the lengths of the cycles.

4. Now ask each group to predict in which year the next cycle of sunspot activity will have its maximum and in which year it will have its minimum.

5. After each group has done this, they should share their results with the class in turn. In their astronomy lab notebooks, students can write down each group's prediction and then see if they can derive a class average. Tell the class that solar astronomers regularly do this kind of activity themselves. They try to predict when the next solar maximum will be and, because the sunspot cycles are not the same length each time, their predictions do not always agree. Ask them to think about why each cycle may not be exactly the same length. (*The reasons for the Sun's activity are very complicated and involve huge areas of the Sun's surface layers—many sectors with their own strong and complex magnetic fields.*)

6. Back in small groups, students can think about cycles in their own lives and experiences and which ones repeat exactly and which ones repeat with big differences. Have them write down their ideas in their astronomy lab notebooks. (*For example, the cycles of our calendar repeat pretty exactly [except for leap years], and the students' birthdays come at the same time*

each year. On the other hand, cycles of the weather are more unpredictable. In some years, winter storms start earlier and in others they start later. While the seasons repeat on a cycle that averages three months, the exact time it gets warmer and colder, or stormier or clearer, may change from year to year. Weather is another example of a complex system, like the seething magnetic activity on the Sun.)

Averaged Sunspot Numbers for the Past Few Decades

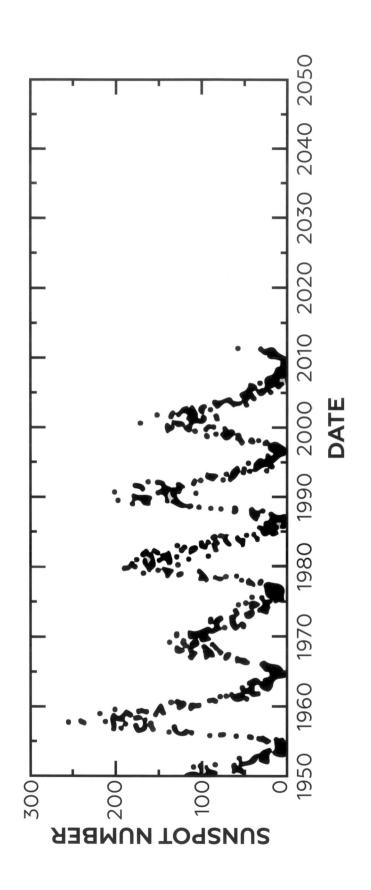

Follow-Up (Extension) Activities
for This Chapter

1. NASA has a detailed activity available online called "Build a Space Weather Action Center." Instructions and downloadable materials can be found at *http://sunearthday.nasa.gov/ swac/gettingstarted*.

2. "Solar Wind Events Tracked to the Source" (part of Space Weather Forecast, a curriculum from the Chabot Space and Science Center) has students examine data about increases in solar wind speed near Earth (gusts of wind) and then study images of the active Sun from a few days before to see if they can find the source of the wind gusts: *http://solar-center.stanford. edu/SID/Distribution/SuperSID/supersid_ v1_1/Doc/SpaceWeatherForecast-v.070507. pdf#page=71*.

3. "Sunspot Races" is an activity suggested by the Stanford Solar Center. After students look at images of the Sun and measure the rotation rate (as in Experience 3.5, "How Fast Does the Sun Rotate") they could look at images from the next few days and then try to predict when the larger sunspot groups that are just disappearing on one side of the Sun (and going behind the Sun) will reappear on the Sun's other limb (side).

   You can get other images of the Sun like the ones we use in this chapter at *http:// sohodata.nascom.nasa.gov/cgi-bin/data_ query*. Note that the images we show are from the instrument HMI Continuum. Our images were at a resolution of 1,024. Where you are asked for the start date, you put the date in the format YEARMONTHDAY (e.g., June 15, 2014 = 201406015). You then get a list of images for different hours of that day,

and you can click on any of them to see the sunspots that day. We used the images from 1500 hours each day.

   Other sources for good images of the Sun are given in section 9 of the Procedure for Experience 3.8, "The Multicolored Sun." Have students follow the progress of the sunspots day by day on these images and then make their predictions. You could have Starburst candy for the winners.

4. A more detailed activity in which students try to see (as Galileo did) whether sunspots are on the Sun or could instead be dark objects in orbit around it, can be found at the Stanford Solar Center: *http://solar-center.stanford. edu/activities/galileo-sunspots.html*.

5. A brief activity from NASA's Goddard Space Flight Center allows students to compare sunspot activity on our star with similar activity on several nearby stars: *http://image. gsfc.nasa.gov/poetry/activity/s3.pdf*.

6. Several smartphones and tablet applications can tell you and your students more about what's happening on the Sun in real time:

   a. NASA's 3D Sun: *http://3dsun.org*

   b. Solaris Alpha: *https://play.google. com/store/apps/details?id=com. tomoreilly.solarisalpha*

   c. NASA Space Weather: *https:// itunes.apple.com/us/app/ nasa-space-weather/ id422621403?mt=8*

   d. Solar Monitor Pro: *www.solarmonitor.eu*

Video Connections

- *Out There: Raining Fire.* Nice overview and introduction to the Sun by science reporter Dennis Overbye of the *New York Times* (3 min.): *www.nytimes.com/video/science/100000003489464/out-there-raining-fire.html?emc=eta1*

- *What Happens on the Sun Doesn't Stay on the Sun.* A video from the National Oceanic and Atmospheric Administration that introduces to the Sun as well as space weather, its effects, and how we monitor it (5 min.): *www.youtube.com/watch?v=bg_gD2-ujCk*

- *Sun Storms.* A video from the Starry Night company about storms from the Sun now and in the past (5 min.): *www.livescience.com/11754-Sun-storms-havoc-electronic-world.html*

- *Journey Into the Sun.* This 2010 KQED Quest TV program is mostly about the Solar Dynamics Observatory spacecraft, its launch, and its capabilities, but the video includes good general information on how the Sun works (12 min.): *www.youtube.com/watch?v=fqKFQ7z0Nuk*

- *Space Weather Impacts Videos.* Four videos from the National Weather Service and National Oceanic and Atmospheric Administration (3–5 min.): *www.swpc.noaa.gov/content/education-and-outreach*; (also on YouTube at *www.youtube.com/playlist?list=PLBdd8cMH5jFmvVR2sZubIUzBO6JI0PvxO*)

- *NASA/SDO: Three Years in Three Minutes—With Expert Commentary.* A video from NASA's Goddard Space Flight Center of three years of observations of the Sun by the Solar Dynamics Observatory made into a sped-up movie, with commentary by solar physicist Alex Young (5 min.): *www.youtube.com/watch?v=QaCG0wAjJSY&src_*

- *Sunspot Group AR 2339 Crosses the Sun.* An animation with music of Solar Dynamics Observatory images of an especially large sunspot group going across the Sun's face (1 min.): *http://apod.nasa.gov/apod/ap150629.html*

Math Connections

- *Solar Math,* published by NASA, is an entire book of free math examples based on science related to the Sun. Written by Dr. Sten Odenwald and his staff, these problems cover a wide range of topics in both science and math. Check out the entire book if you want to find good connections between the science and the math your students are learning. You can find the book at *www.nasa.gov/audience/foreducators/topnav/materials/listbytype/Solar_Math.html.*

  - Dr. Odenwald continues to pose additional math problems each year related to space and astronomy. You can find all the collections at *http://spacemath.gsfc.nasa.gov/books.html* or use the menu bars at the top to select problems by grade level or other criteria.

- One of our favorite math problems from the series is "An Interplanetary Shock Wave":

On November 8, 2000, the Sun ejected a billion-ton cloud of gas called a coronal mass ejection or CME. On November 12, the CME collided with Earth and produced a brilliant aurora detected from space by the IMAGE satellite. On December 8, the Hubble Space Telescope detected an especially bright aurora on Saturn.

During the period from November to December, 2000, the Earth and Saturn were almost lined up with each other. Assuming that the two planets were located on a straight line drawn from the Sun to Saturn, with distances from the Sun of 150 million and 1.43 billion km, respectively, students should answer the questions below:

    a. How many days did the CME take to reach Earth and Saturn?

    b. What was the average speed of the CME in its journey between the Sun and Earth in millions of km per hour?

    c. What was the average speed of the CME in its journey between Earth and Saturn in millions of km per hour?

    d. Did the CME accelerate or decelerate as it traveled from the Sun to Saturn?

You might also ask the students to convert the speeds from kilometers per hour to miles per hour and then compare the speeds they calculate to the speed of a car or the speed of a jet airplane on Earth. (Odenwald, S., and D. Janney. n.d. *Space Math-II*. NASA Goddard Space Flight Center: Greenbelt, MD. 11–12. *www.nasa.gov/pdf/280167main_Space_Math_II.pdf*)

Here's another good set of problems:

The radius of the Sun (measured at the photosphere) is about 7×10^5 km.

- How many meters is the Sun's radius? (Answer: 7×10^8 m)

- What is the volume of the Sun? (Answer: 1.4×10^{27} m³)

The radius of the Earth is 6,371 km.

- How many meters is that? (Answer: 6.4×10^6 m)

- What is the volume of the Earth? (Answer: 1.08×10^{21} m³)

- How many Earths fit into the equator of the Sun? (Answer: about 10^9)

- How many Earths fit into the volume of the Sun? (Answer about 1.3×10^6, or 1.3 million, or you could just say "a lot!")

- The Sunspot Anatomy page from the National Solar Observatory poses a series of math questions converting angular measure on the sky to linear measure and estimating the size of a big sunspot. It is available at *www.as.utexas.edu/mcdonald/scope/poster/sunspots.pdf*.

- Suntrek is an online activity booklet produced in England in cooperation with Deborah Scherrer of the Stanford Solar Center. It contains discussions of and math problems about solar images, space weather, and the spectrum of the Sun at the high school level. You can find it at *http:// solar-center.stanford.edu/activities/ Suntrek/Suntrek-Solar-Data.pdf*.

- *Electromagnetic Math* is a whole separate NASA space math booklet devoted to problems delving into the electromagnetic spectrum. You can download it free at *www.nasa.gov/audience/foreducators/ topnav/materials/listbytype/ Electromagnetic_Math.html*.

Literacy Connections

- A short introduction for younger students from NASA's Space Place:
    NASA. What is space weather? *http:// spaceplace.nasa.gov/spaceweather/en*.

- A booklet of illustrations to help older students understand space weather:
    U.S. Department of Commerce, National Oceanic and Atmospheric Administration, and National Weather Service. *Space Weather: Storms on the Sun. www.swpc. noaa.gov/sites/default/files/images/u33/ swx_booklet.pdf*.

- A page of basic solar information, including a top 10 facts about the Sun section:
    NASA. Sun for kids. *http://spaceplace. nasa.gov/external/http://www.nasa.gov/ vision/universe/solarsystem/Sun_for_ kids_main.html*.

- Find out more about the Sun from the High Altitude Observatory:
    High Altitude Observatory. Questions and answers about the Sun. *http:// www2.hao.ucar.edu/Education/ questions-and-answers-about-sun*.

- A document with questions and answers about auroras from NASA:
    NASA. Auroras. *http://pwg.gsfc.nasa. gov/polar/EPO/auroral_poster/ aurora_all.pdf*.

- A NASA news report about the Carrington event and modern solar storms:
    NASA. A super solar flare. *http://science. nasa.gov/science-news/science-at-nasa/ 2008/06may_carringtonflare*.

- A series of web readings, video, and colorful booklet to print out:
    NASA. Tour of the electromagnetic spectrum. *http://missionscience.nasa.gov/ ems/index.html*.

Cross-Curricular Connections

- A famous science fiction story about the solar wind, first published in *Boy's Life*, is Arthur C. Clarke's "The Wind from the Sun." It is available free online at *www. baenebooks.com/chapters/ 9781625792112/9781625792112___3. htm*. Students should read the story and then write about it in their astronomy lab notebooks. What correct or incorrect science do they recognize in the story? Could what happens in the story ever happen in the future? If they have time, they can also do research on the internet about the solar sails that have been launched so far and what has happened to them.

One group trying to get solar sails launched and tested is the nonprofit Planetary Society. Students can read more about their project at *http://sail. planetary.org*.

- After students are finished with their experiences and their study of the Sun and space weather, give them the assignment to write a short story about the effect of space weather on a Moon colony of the future. Students can imagine that, decades from now, there is a domed city on the Moon, with an atmosphere and heat under the dome, where people live and work. Suddenly, a flare is observed and a CME is seen by satellites monitoring the Sun to be aimed in the direction of the Earth and the Moon. The story should describe what would happen next and imagine some ways the Moon colony might prepare for a solar storm and some ways in which they might not be able to prepare.

- Students can research the Carrington event of 1859, when a large CME from the Sun came directly toward our planet. Can they find any reports from people at the time (this is a good task for web and library research, perhaps with help from the school librarian)? Then they could summarize what happened in 1859, either as an essay or as a news report (imagining that there was cable news back then.) What would the average person reading the newspaper in those days have learned about the event? (If they have trouble finding such reports on their own, they can see, for example,

www.astronomycafe.net/NewsStorms. html.)

Students could also research events since that time. A remarkable book full of newspaper and other accounts of solar storms, that you might ask your library to get, is astronomer Sten Odenwald's *Solar Storms: 2000 Years of Human Calamity*. His website, Solar Storms (*www. solarstorms.org*), has links to some of these reports. The full book is available on Amazon at *www.amazon.com/ Solar-Storms-years-human-calamity/ dp/1505941466*.

- In 1970, composer Charles Dodge wrote an electronic piece of music for which he sonified (turned into notes) the index for auroral activity for 1961. He used something called the Kp index, which is related to the aa index we used in Experience 3.7. As he said in the liner notes for the published album, "The succession of notes in the music corresponds to the natural succession of the *Kp* indices for the year 1961. … The musical interpretation consists of setting up a correlation between the level of the *Kp* reading and the pitch of the note (in a diatonic collection over four octaves), and compressing the 2,920 readings for the year into … musical time." Your students can listen to the piece at *www.youtube. com/watch?v=j5MHsnc67yw* and learn a bit more at *http://info.umkc.edu/ specialcollections/archives/1056*.

This is just one of a series of modern musical pieces in which the sounds are based on astronomical data. Another is the *Supernova Sonata*, music based on

the characteristics of stars that explode at the end of their lives: *www.astro.uvic.ca/~alexhp/new/supernova_sonata.html*.)

Your students could discuss whether this is a good way to make music or not.

- Students interested in computer applications might want to check out the Sun apps recommended in item 6 of the Follow-Up (Extension) Activities for This Chapter on page 263.

- For more cross-curricular ideas related to the Sun, see the book by Richard Cohen, *Chasing the Sun: The Epic Story of the Star That Gives us Life*. A journalist and professor of creative writing, Cohen explores the Sun in mythology, art, literature, psychology, and science.

Resources for Teachers

BOOKS AND ARTICLES

- A likeable introduction for beginners.
  Golub, L., and J. M. Pasachoff. 2002. *Nearest star: The surprising science of the Sun*. Boston: Harvard University Press.

- A popular guide to the Sun's activity and its effects on Earth.
  Carlowicz, M., and R. Lopez. 2002. *Storms from the Sun: The emerging science of space weather*. Washington, DC: Joseph Henry Press.

- A beautifully illustrated introduction to the Sun for beginners by a Norwegian solar astronomer.

Brekke, P. 2012. *Our explosive Sun: A visual feast of our source of light and life*. New York: Springer Verlag.

- A nice introduction by an astronomer to the Sun and space weather:
  Odenwald, S. 2012. *The 23 cycle: Learning to live with a stormy star*. New York: Columbia University Press.

- An up-to-date review of how events on the Sun can hurt our civilization:
  Berman, B. 2013. How solar storms could shut down Earth. *Astronomy Magazine* (September): 22.

- A brief introduction to what there is to see on the Sun and how to see it:
  Bakich, M. E. 2008. How to observe the Sun. *Astronomy Magazine* (April): 64.

WEBSITES

- A primer on space weather:
  National Oceanic and Atmospheric Administration. A profile of space weather. *www.swpc.noaa.gov/sites/default/files/images/u33/primer_2010_new.pdf*.

- Dr. Sten Odenwald's Solar Storms site:
  Odenwald, S. Space weather. Solar Storms. *www.solarstorms.org*.

- An excellent site with information for students and teachers:
  Stanford Solar Center. *http://solar-center.stanford.edu*.

- A site for beginners:
  High Altitude Observatory. The Sun: A pictorial introduction. *https://www2.hao.ucar.edu/Education/sun-pictorial-introduction*.

- A National Oceanic and Atmospheric Administration page that includes primers, videos, a curriculum and training modules: Space Weather Prediction Center. Education and outreach. *www.swpc.noaa. gov/content/education-and-outreach.*

- This site provides a nice primer for students on how the Earth and the Sun interact: Space Science Institute. Space weather center. *www.spaceweathercenter.org.*

- Videos, scientist profiles, a research challenge related to the active Sun from the PBS science program: Nova Labs. The Sun lab. PBS. *www.pbs. org/wgbh/nova/labs/lab/Sun.*

- The Cool Cosmos gallery shows you objects on Earth and in the universe in visible light and in infrared: Cool Cosmos. IPAC. *http://coolcosmos. ipac.caltech.edu.*

4

The Sun, the Moon, and the Earth Together

PHASES, ECLIPSES, AND MORE

Chapters 1 and 2 focused on important cycles of time produced by the rotation of the Earth on its axis and the revolution of Earth in its orbit around the Sun. These determine how we measure two key periods of our lives: the day and the year. This chapter adds a third cycle of importance in astronomy (and human culture): the revolution of the Moon around the Earth. The Moon's cyclical motion is not only connected to our notion of a month, but it also produces the changing phases of the Moon and explains why we have solar and lunar eclipses.

FIGURE 4.1

An image of the Moon taken with the NASA Galileo spacecraft

Note the lighter rays coming from the Tycho crater at the bottom of the Moon's disk.

4

Learning Goals of the Chapter

After doing these activities, students will understand the following:

1. As the Moon orbits the Earth, its position relative to the Sun changes and produces the different lunar phases we see.

2. The lunar phases have a predictable pattern over about a month's time, going from new Moon to waxing crescent, to first quarter, to waxing gibbous, to full Moon, to waning gibbous, to third quarter, to waning crescent, and back to new Moon.

3. Solar eclipses can only occur when the Moon is in its new moon phase.

4. Lunar eclipses can only occur when the Moon is in its full Moon phase.

5. Because the orbit of the Moon is tilted with respect to the orbit of the Earth, lunar and solar eclipses do not occur each month as one might expect but instead occur in pairs of solar and lunar eclipses around every six months.

6. While people on half of the Earth can see a total lunar eclipse, only people in a narrow path on the Earth can see a total solar eclipse.

7. Because the Moon takes the same amount of time to rotate as it takes to revolve around the Earth, we on Earth only see one side of the Moon.

Overview of Student Experiences

Teacher note: For maximum effectiveness, the first three experiences in this chapter should be done significantly in advance of when you do the experiences that follow. This will allow students to have 10 to 30 days to observe the Moon on their own before doing the later experiences in this chapter. This might mean interrupting another science unit with the first three activities, but we assure you that the effectiveness of keeping a Moon observation journal far outweighs the short-term challenge of starting this chapter before finishing another science topic.

Alternatively, you can begin your unit with the first three experiences and then move to other chapters of this book, allowing for a 10- to 30-day gap between the time you do the first three experiences in this chapter and when you do the rest.

ENGAGE EXPERIENCES

This chapter starts with what appears to be a simple challenge for the students: putting a series of lunar photographs in order by the shape of the Moon as it would appear if they observed it over a number of weeks. Because the students want to know if their answers are correct, this experience "hooks" them into observing the Moon, which ultimately leads to a deeper understanding of lunar phases, the connection of the lunar phase cycle to our concept of the month, and the cause of solar and lunar eclipses.

- **4.1. Predicting What the Moon Will Look Like:** Students examine six photographs of the Moon and predict the order of the different phases.

- **4.2. What Do We Think We Know?** Students discuss and record their thoughts regarding the following questions: (1) What are the different phases of the Moon, and how do the phases change throughout the month and year? (2) What causes the different phases? (3) What are solar and lunar eclipses, and what causes them?

EXPLORE EXPERIENCE

Students make their own observations of the Moon, which they will need for the *explain* experiences, including the time they see the Moon, the phases they see, the position of the Moon in the sky, and key features on the lunar surface (e.g., craters, maria).

- **4.3. Observing the Moon:** Students begin making basic observations of the Moon when it is out during the school day, followed by observations in the evening sky from their homes. They use the data collected over a number of days to conclude how the Moon's phases and its position in the sky change over time.

EXPLAIN EXPERIENCES

Building from their lunar observations, students create a model with their heads (the Earth), a bare lightbulb (the Sun), and a Styrofoam ball (the Moon) to understand what causes the lunar phases and solar and lunar eclipses.

- **4.4. Modeling the Moon:** This activity builds on the modeling activities from Chapters 1 and 2 in which the students' heads are the Earth and a lightbulb in the front of the room is the Sun. Each student receives a model Moon (a small Styrofoam ball attached to a pencil) that they use to explore the sequence of lunar phases they see as the Moon orbits the Earth.

- **4.5. Modeling Eclipses:** After students understand where the Moon has to be in its orbit to see each phase, the modeling process continues as they explore where the Moon has to be to produce solar and lunar eclipses.

ELABORATE EXPERIENCES

Students engage in additional experiences that reinforce and build a deeper understanding of the core ideas and science practices regarding the Moon's motions and phases and the timing of and explanation for eclipses.

- **4.6. How Often Do Eclipses Occur?** The teacher uses hula hoops to model the Moon's orbit around the Earth and the Earth's orbit around the Sun to help students understand why there are typically two solar eclipses and two lunar eclipses each year.

- **4.7. Why Do People Spend $10,000 to See a Total Solar Eclipse?** Students use the Earth–Moon–Sun model to discover that the shadow of the Moon only covers a small portion of the Earth (the student's head), leading to the realization that people who want to see a solar eclipse often need to travel to elsewhere in the world to do so. Students then play with the model to discover that this is not true for a lunar eclipse: The entire half of the Earth facing toward the Moon will see a lunar eclipse.

4.8. Does the Moon Rotate? Many people believe that the Moon does not rotate because we always see the same side of the Moon facing the Earth. Students use a model in which students' heads are the Moon and Earth to determine whether or not the Moon rotates.

4.9. What Do Eclipses Look Like From a Space Colony on the Moon? Students use their Earth–Moon–Sun model to predict what they would see during a solar eclipse if they were astronauts at a Moon base. They then use internet resources to see simulations of this as well as observations made from the International Space Station that show what astronauts there see during solar and lunar eclipses.

EVALUATE EXPERIENCES

These experiences allow both students and teachers to assess how well the students understand the key ideas presented in the chapter.

4.10. Lunar Phases Revisited: Students are given a new set of six photographs of the Moon and are asked to repeat the process of predicting the order in which they would see the shape of the Moon if they observed it over a month's time.

4.11. What Causes Lunar Phases and Eclipses? Students are asked to label a diagram that shows the Earth, Sun, and Moon to indicate where the Moon is in its orbit to produce each phase and to note where the Moon must be in its orbit to produce solar and lunar eclipses.

Recommended Teaching Time for Each Experience

Table 4.1 shows the amount of time needed for each experience.

Connecting With Standards

Table 4.2 (pp. 276–277) shows the standards covered by the experiences in this chapter. Chapter 2 dealt with any phenomena associated with seasons and the seasonal changes related to the apparent motion of the Sun in the sky, so this chapter emphasizes the cyclic patterns of lunar phases and eclipses. Additionally, this table does not give *Common Core State Standards,* but gives general connections to the language arts and mathematics standards.

TABLE 4.1

Recommended teaching time for each experience

Experience	Time
Engage experiences	
4.1. Predicting What the Moon Will Look Like	**45 minutes**
4.2. What Do We Think We Know?	**45 minutes**
Explore experiences	
4.3. Observing the Moon	**45 minutes** for introduction of experience, followed by **10 minutes each day** for follow-up observations; requires students to make observations outside of class time and a number of days with clear weather; summing up of what students learn takes another **45 minutes**
Explain experiences	
4.4. Modeling the Moon	**Two 35-minute** periods
4.5. Modeling Eclipses	**45 minutes**
Elaborate experiences	
4.6. How Often Do Eclipses Occur?	**45 minutes**
4.7. Why Do People Spend $10,000 to See a Total Solar Eclipse?	**45 minutes**
4.8. Does the Moon Rotate?	**45 minutes**
4.9. What Do Eclipses Look Like from a Space Colony on the Moon?	**45 minutes**
Evaluate experiences	
4.10. Lunar Phases Revisited	**45 minutes**
4.11. What Causes Lunar Phases and Eclipses?	**30 minutes**

TABLE 4.2

Chapter 4 *Next Generation Science Standards* and *Common Core State Standards* connections

Performance expectations	• MS-ESS1-1: Develop and use a model of the Earth–Sun–Moon system to describe the cyclic patterns of lunar phases, eclipses of the sun and moon, and seasons.
Disciplinary core ideas	• MS-ESS1.A: (The Universe and Its Stars) Patterns of the apparent motion of the Sun, the Moon, and stars in the sky can be observed, described, predicted, and explained with models.
Science and engineering practices	• Develop and use a model to describe phenomena (e.g., model of the Earth–Moon–Sun system). • Analyze and interpret data to determine similarities and differences in findings (e.g., daily observations of the lunar phases are compared with observations of others and to the student's own prediction). • Engage in argument from evidence (e.g., discussion of which order of the lunar phases is the most appropriate).
Crosscutting concepts	• Patterns can be used to identify cause-and-effect relationships (e.g., observations of the Moon's phases and modeling of the Earth–Moon–Sun relationship reveals what causes the lunar phases). • Science assumes that objects and events in natural systems occur in consistent patterns that are understandable through measurement and observation (e.g., the pattern of lunar phases is revealed though observations of the Moon). • Systems and system models (e.g. use of Earth–Moon–Sun model to understand the cause of lunar phases and eclipses).

Content Background

THE PHASES OF THE MOON

Earth's natural satellite (which we call the Moon with a capital *M*, to distinguish it from the moons of other planets) is the only object we can see in the sky with just our eyes that changes its appearance on a regular cycle. The differing amounts of light and dark on the face of the Moon are called its phases, and they repeat every 29.5 days with clockwork regularity.

Since the Moon is a world made of cold rock, it produces no light of its own. Moonshine, as the ancient Greeks figured out, is really reflected

Table 4.2 *(continued)*

Connections to the **Common Core State Standards**	• *Writing:* Students write arguments that support claims with logical reasoning and relevant evidence, using accurate, credible sources and demonstrating an understanding of the topic or text. The reasons and evidence are logically organized, including the use of visual displays as appropriate.
	• *Speaking and listening:* Students engage effectively in a range of collaborative discussions (one-on-one, in groups, and teacher-led) with diverse partners, building on others' ideas and expressing their own clearly. Report on a topic or text or present an opinion, sequencing ideas logically and using appropriate facts and relevant, descriptive details (including visual displays as appropriate) to support main ideas or themes.
	• *Reading:* Students quote accurately from a text when explaining what the text says and when drawing inferences from the text. Students determine the meaning of general academic and domain-specific words and phrases in a text relevant to the student's grade level.
	• *Mathematics:* Students recognize and use proportional reasoning to solve real-world and mathematical problems. Students summarize numerical data sets in relation to their context, including reporting the number of observations and describing the nature of the attribute under investigation, including how it was measured and its units of measurement.

sunshine. Therefore, to understand the cycle of the Moon's phases, we must look at the relationship—in the course of a month—between the Earth, the Moon, and the Sun.

For example, when the Sun is shining on the side of the Moon facing the Earth, we on Earth see the Moon lit up with sunshine—what we call a full Moon. On the other hand, when the Sun is shining only on the side of the Moon facing away from the Earth, we see no reflected sunlight from the Moon—what we call new Moon. Between those two phases, we see some part of the side of the Moon facing us lit up and some part dark. It

is this sequence of phases that this chapter helps students to understand (Figure 4.2, p. 278).

THE SURFACE FEATURES OF THE MOON

In a very general way, the surface of the Moon can be divided into two types of terrain, the dark round *maria* and the lighter *highlands* (Figure 4.3, p. 279). The word maria is Latin for "seas," a designation that dates from centuries ago when early Moon observers thought that the shadowy, round regions were quiet, dark oceans. Today we know the maria (pronounced "mar-ee-ah")

FIGURE 4.2

A diagram of the Moon phases showing the full Moon and new Moon

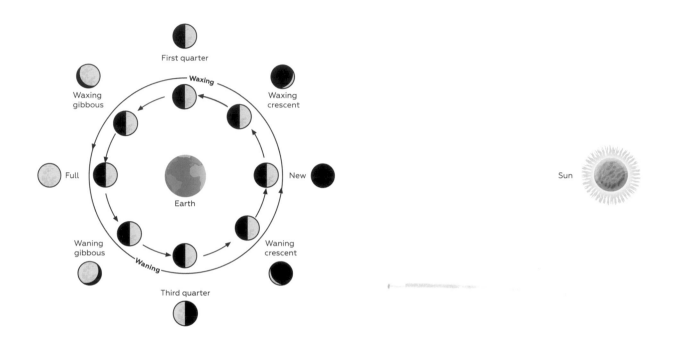

The outer circle of Moon diagrams shows what is visible in the sky from Earth. The inner circle of Moon diagrams shows what would be visible from space, looking down from above the Earth–Moon system.

are where overlapping round impact craters were flooded with darker lava from beneath the Moon's surface. Maria is the plural of mare (pronounced "mah-ray").

The highlands get their name from the fact that they are at a somewhat higher altitude than the maria. It is the contrast of the dark maria with the lighter highlands that has inspired many generations of Moon watchers to imagine that they can see familiar shapes on the side of the Moon facing us, including a man in the Moon, a lady's face, two children, and animals (Figure 4.4).

The Moon is full of *craters*—depressions made when chunks of rock or ice hit the Moon and explode with the energy of impact. Such explosions carve out a round hole in the surface of the Moon roughly 10 to 20 times the size of the object that exploded. Younger craters tend to be brighter, as their impact exposes material that has not yet been covered by the darker Moon dust that blankets so much of our natural satellite. You can see some large, young craters in Figure 4.3, including Tycho and Copernicus (also see the "Lunar Map" on p. 300). On good photographs of the Moon, you

FIGURE 4.3

A photo of the Moon with major features labeled (see p. 300 for a larger version)

FIGURE 4.4

People imagine they can see various faces and animals, such as a rabbit, in the shapes on the Moon.

can see that the shapes of the dark maria all consist of combinations of big craters that were made long ago when giant chunks of rock were still common in the inner solar system.

Depending on the size and nature of the explosion that forms a crater, good-sized chunks of exploded rock can be thrown sideways away from the site of the explosion. When these pieces fall, approaching the surface and scraping along it before coming to a rest, they leave lighter streaks along their paths. Called *rays*, some of these features on the Moon can extend dozens or even hundreds of miles from the original crater, but they all point back to the original explosion that produced them. Rays from the crater Tycho are easily seen on photos of the Moon that show good detail (Figure 4.5).

FIGURE 4.5

Rays from a crater on the Moon

On this image of the Moon taken by the Clementine spacecraft, the bright crater Tycho has rays coming out of it in many directions.

FIGURE 4.6

An illustration of the Moon's orbit relative to the Earth and Sun

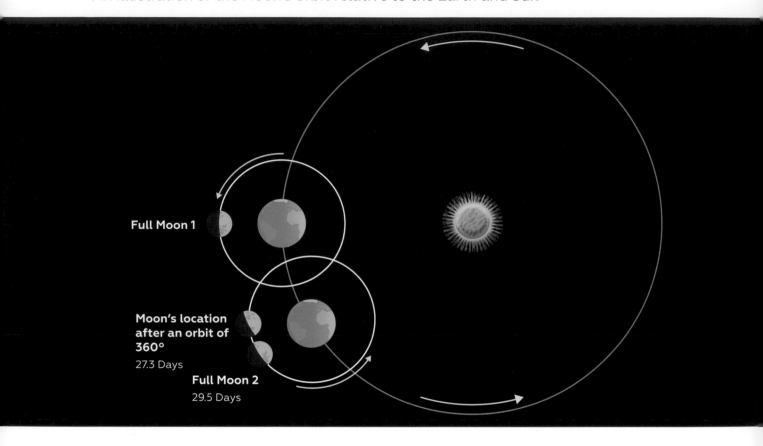

Full Moon 1

Moon's location
after an orbit of
360°
27.3 Days

Full Moon 2
29.5 Days

It takes two extra days for the Moon to go from full Moon to full Moon compared with the time it takes to complete one full orbit (360°) of the Earth because the Earth is also orbiting the Sun. Thus, the Moon has to revolve more than 360° to get back in line with the Earth and Sun, which is required to see a full Moon.

THE MONTH AND THE MOTIONS THAT DEFINE IT

The Moon takes 27.3 days to orbit (revolve around) the Earth—that is, to return to the same place relative to the stars. However, the Earth–Moon system is also revolving around the Sun during that time, which means that the relative position of the three bodies is slowly changing. This is why the phases of the Moon take about 29.5 days to repeat (Figure 4.6). It is this cycle of repeating phases that has given its length to the unit of time we call the month, a term that is based on the word Moon.

The cycle of the Moon's phases over 29.5 (roughly 30) days has been one of the fundamental elements of timekeeping in human history. Since the Moon is easy to see and remember, it is simple to keep track of the cycle. Many cultures in earlier millennia and centuries kept an annual calendar based on the Moon's cycle and not the Earth's yearly trip around the Sun. Since the 29.5-day month does not go evenly into the 365.25-day solar year, adjustments had to be made to the length of each month to get 12 months to fit into 365 days. In our western calendar, we adjust (somewhat awkwardly) by making some months have more days and one month (February) have fewer days than 30.

In other traditions, a calendar based on the lunar cycle dominated, which is why certain holidays (Easter, Hannukah, Ramadan) come on different calendar days in different years.

THE MOON'S ROTATION

In addition to revolving around the Earth, the Moon also spins on an axis. It takes the same amount of time to turn as it takes to orbit the Earth. This "synchronous rotation" means that the Moon turns around the Earth at the same time as it turns around itself. Thus, it always keeps roughly the same face toward our planet—which is why there is a near side and a far side to the Moon.

Such synchronous rotation is not confined to our Moon. Jupiter's four giant moons, Io, Europa, Ganymede, and Callisto, also show this connection between their time for orbiting and spinning, as does Pluto's large moon Charon. When a planet and a moon can exchange rotational energy (through some process like the ocean tides that the Moon pulls on Earth), their tendency is to make rotational motions equal to or small ratios of one another. (Long ago, after it formed, the Moon was molten, and the differences in the Earth's pull on its front and back side deformed it into a shape that was not entirely spherical. This asymmetrical shape of the Moon allows the

FIGURE 4.7

A diagram of a solar eclipse (not to scale)

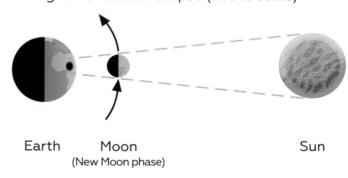

Earth   Moon   Sun
        (New Moon phase)

Earth's gravity to pull on the Moon in a way that slowly changes its motion).

UNDERSTANDING SOLAR ECLIPSES

A solar eclipse happens when the Sun, Moon, and Earth are lined up, and the Moon covers some or all of the Sun as seen from Earth. The Moon must be in its new Moon phase for a solar eclipse to be possible (Figure 4.7).

It is a remarkable coincidence that, in our present epoch, the Sun as seen from Earth and the Moon as seen from Earth happen to be the same angular size in the sky. Therefore, the disk of the Moon can exactly cover up the disk of the Sun in what is called a *total solar eclipse* (Figure 4.8a, p. 282). Eclipses of the Sun are not always total, however. If the Moon is a bit farther away from us in its not-entirely circular orbit, then it doesn't quite manage to cover up the outermost ring of the Sun, leading to an *annular eclipse* (Figure 4.8b). And if the Moon is not lined up precisely with the Sun, we get only a piece of the Sun covered up in what is called a *partial eclipse* (Figure 4.8c).

FIGURE 4.8

Photographs of a (a) total, (b) annular, and (c) partial solar eclipse

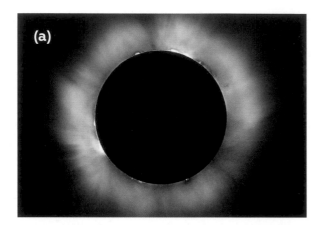

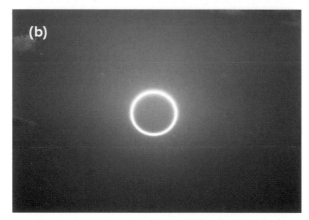

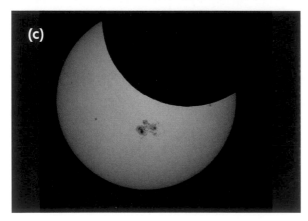

During a total eclipse of the Sun, the Moon's darkest shadow makes a spot on the Earth's surface, roughly 60–160 mi. wide, that moves across the Earth as the Moon and Earth move relative to each other, producing what is called the *path of totality*. If you are standing in that path, you will see the Moon slowly cover the entire Sun until the faint outer atmosphere of the Sun (the corona) appears like a ring of faint light around the dark disk of the Moon (see the corona around the edge of the Sun in Figure 4.8a). Because the Sun is completely covered for a few minutes during the total solar eclipse, the sky goes dark and, if it is not cloudy, the stars come out. Animals become confused by what is happening. For example, birds become quiet and begin to roost as if night were coming. This experience of darkness in the midst of day is quite beautiful and awe-inspiring. People pay thousands of dollars to go to remote parts of our planet just to stand in the Moon's shadow and witness such a thing. But the total phase is brief—in less than five minutes (usually much less), the Moon starts to uncover the Sun and totality is over (Figure 4.9).

UNDERSTANDING LUNAR ECLIPSES

A lunar eclipse, on the other hand, happens when the Sun, Earth, and Moon are lined up, and the Earth casts its shadow on the full Moon. Lunar eclipses can only occur during the full Moon phase of the Moon (Figure 4.10). The Earth's shadow on the Moon is much bigger than the smaller Moon's shadow on the Earth. Thus, the Earth's shadow can cover the entire surface of the Moon during a total lunar eclipse, and we can see the entire face of the Moon go slowly dark (Figure 4.11).

As the Earth's shadow moves across the bright full Moon, everyone on the side of the Earth facing the Moon can see the eclipse. Thus, many more people get to witness a total eclipse of the Moon than a total eclipse of the Sun. No one travels great distances to see a lunar eclipse; people can stay where they are and nature will bring a lunar eclipse their way soon enough.

If the three bodies are not entirely lined up, it's possible to have a partial eclipse of the Moon, in which only part of the full Moon is covered. As with solar eclipses, the partial lunar eclipse is not quite as spectacular as the total one, but it's still interesting for students to try to see.

During a total lunar eclipse, the Moon does not go completely dark but turns a reddish-brown color. This is the result of sunlight refracting (bending) onto the Moon in the Earth's thick atmosphere. Much like we see at sunset, the red color from the Sun gets through the Earth's atmosphere

FIGURE 4.9

The path of the August 2017 total eclipse of the Sun

FIGURE 4.10

A diagram of a lunar eclipse (not to scale)

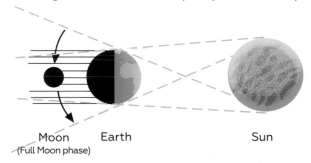

Moon   Earth   Sun
(Full Moon phase)

FIGURE 4.11

Stages in a total lunar eclipse

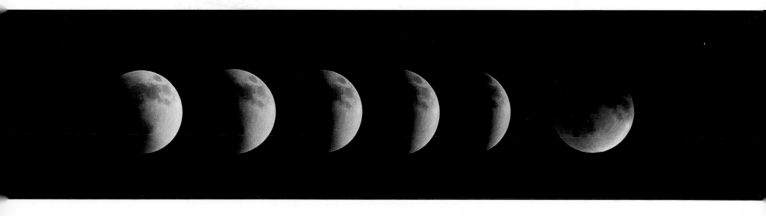

FIGURE 4.12

A diagram showing how light diffracts during a total lunar eclipse

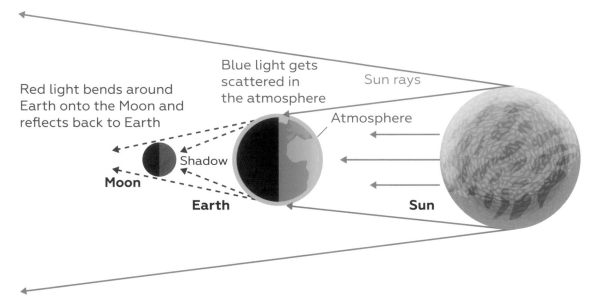

A total lunar eclipse occurs when the Moon enters the Earth's shadow. If the Earth did not have an atmosphere, the Moon would go completely black, but Earth's atmosphere acts like a lens to bend the sunlight into the Earth's shadow. The atmosphere also scatters the shorter wavelengths of light (blue and green) more and the longer wavelengths (orange and red) less. This means mostly orange and red light get through to continue toward the Moon. When the orange and red light reaches the Moon, these colors are reflected back to the Earth, so we see the Moon not as being completely dark but as having a reddish hue. (Image not to scale.)

while the other colors are scattered (Figure 4.12). The exact color of the Moon depends on how polluted our atmosphere is with small particles of dust, dirt, or ash. Lunar eclipses are always more colorful after large volcanic eruptions, for example.

WHY ECLIPSES DON'T HAPPEN EVERY MONTH

Since there is a full Moon and new Moon every month, students often wonder why we don't have a lunar and solar eclipse every month. The key to why eclipses don't happen every month is that the plane of the Moon's orbit is not exactly aligned with the plane of the Earth's orbit around the Sun. The Sun appears to travel along what is called the *ecliptic*, and the Moon's orbit is inclined about 5° relative to this (Figure 4.13). Therefore, in most months, we see the Moon a little bit above or below the position where we see the Sun in the sky. Only twice a year, when the two orbits intersect, can the Sun, Moon, and Earth line up in such a way as to make an eclipse possible. During these roughly twice yearly "eclipse seasons," we typically have a pair of eclipses, one during the full Moon and the other during the new Moon, followed by another pair of solar and lunar eclipses about six months later.

For a list of upcoming solar and lunar eclipses, see Tables 4.3 and 4.4 on pages 286 and 287 or visit *www.nsta.org/solarscience*.

SAFE ECLIPSE VIEWING

Important warning: Viewing all or part of the Sun without proper protection is dangerous—except for the few minutes during the total phase of an eclipse when one can look at the Sun with unprotected eyes and binoculars!

The Sun's visible (and invisible) rays can cause serious damage to the sensitive tissues of the eye, often without one being immediately aware of it! Normally, our common sense protects us from looking directly at the Sun for more than a second. But during an eclipse, astronomical enthusiasm can overwhelm common sense, and people can wind up staring at the Sun for too long. Make sure you have something with you to protect your eyes before the eclipse becomes total or if you are only seeing the partial eclipse.

The few minutes of total eclipse (when the Sun is completely covered) ARE safe, but any time that even a small piece of the bright Sun shows,

FIGURE 4.13

The paths of the Sun and the Moon in the sky

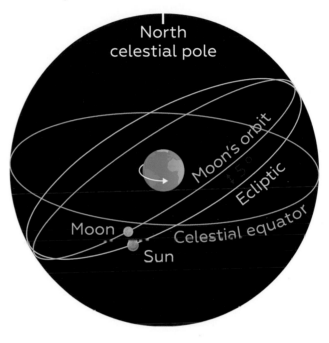

your eyes are in danger. See Experience 3.3, "Safe Solar Viewing," for detailed instructions on how to use binoculars to project an image of the Sun that the whole class can look at together. Experience 3.10 (pp. 250–256) describes some other safe ways to look at the Sun.

TABLE 4.3

Future total and annular solar eclipses

Date	Type of eclipse	Location on Earth ●
Sep. 1, 2016	Annular	South Atlantic Ocean, central Africa, Madagascar, Indian Ocean
Feb. 26, 2017	Annular	Southwest Africa, southern tip of South America
Aug. 21, 2017	Total	United States (Oregon to South Carolina) and oceans on either side
July 2, 2019	Total	Southwest South America, Pacific Ocean
Dec. 26, 2019	Annular	Saudi Arabia, southern India, Malaysia
June 21, 2020	Annular	(very short) central Africa, Pakistan, India, China
Dec. 14, 2020	Total	Chile, Argentina, and oceans on either side
June 10, 2021	Annular	Northern Canada, Greenland
Dec. 4, 2021	Total	Only in Antarctica
April 20, 2023	Total ○	Mostly Indian Ocean and Pacific Ocean, plus Indonesia
Oct. 14, 2023	Annular	Oregon, Nevada, Utah, New Mexico, Texas, Central America, Colombia, Brazil
April 8, 2024	Total	Northern Mexico, United States (Texas to Maine), southeastern Canada, oceans on either side
Oct. 2, 2024	Annular	Southern Chile, southern Argentina, oceans on either side
Feb. 17, 2026	Annular	Antarctica only
Aug. 12, 2026	Total	Greenland, Iceland, Spain
Feb. 6, 2027	Annular	South Pacific, Argentina, Chile, Uruguay, southern Atlantic Ocean
Aug. 2, 2027	Total	Spain, Morocco, Egypt, Saudi Arabia, Yemen, Arabian Sea
Jan. 26, 2028	Annular	Ecuador, Peru, Brazil, northern Atlantic Ocean, Portugal, Spain
July 22, 2028	Total	Indian Ocean, Australia, New Zealand, southern Pacific Ocean

● Remember that a total or annular eclipse is only visible on a narrow track. The same eclipse will be partial over a much larger area, but partial eclipses are not as spectacular as total ones.

○ This is a so-called hybrid eclipse, which is total in some places and annular in others

NATIONAL SCIENCE TEACHERS ASSOCIATION

TABLE 4.4

Future total lunar eclipses

Note: Because partial and penumbral lunar eclipses are not that spectacular, we list only the eclipses that become total (when the Moon is completely within the Earth's dark shadow).

Date	Type of eclipse	Location on Earth
Jan. 31, 2018	Total	Asia, Australia, western North America
July 27, 2018	Total	South America, Asia, Africa, Australia, Indian Ocean
Jan. 21, 2019	Total	North America, South America, western Africa, western Europe
May 26, 2021	Total	Eastern Asia, Australia, Pacific Ocean, western North America, western South America
May 16, 2022	Total	North America, South America, Europe, Africa
Nov. 8, 2022	Total	Asia, Australia, Pacific Ocean, North America, South America
March 14, 2025	Total	Pacific Ocean, North America, South America, Atlantic Ocean, western Europe, western Africa
Sep. 7, 2025	Total	Europe, Africa, Asia, Australia, Indian Ocean
March 3, 2026	Total	Eastern Asia, Australia, Pacific Ocean, North America, Central America
June 26, 2029	Total	Eastern North America, South America, Atlantic Ocean, western Europe, western Africa
Dec. 20, 2029	Total	Eastern North America, Eastern South America, Atlantic Ocean, Europe, Africa, Asia

4.1
ENGAGE

EXPERIENCE 4.1

Predicting What the Moon Will Look Like

Overall Concept

Students examine and predict the order of six photographs of the Moon that show different phases. Their predictions are then compared to the actual observations students make over the next 10 to 30 days as part of *explore* Experience 4.3, "Observing the Moon."

Objectives

Students will

1. use photos of the Moon to predict the sequence of the Moon's phases based on their prior knowledge,

2. recognize that the Moon's overall appearance changes on a regular cycle,

3. question how and why the Moon's appearance changes, and

4. identify a number of features on the Moon's surface.

   Teacher note: Many students—and teachers—think of the Moon as having eight phases (see, for example, Figure 4.14, p. 297). There is not a specific number of phases, since the Moon goes through a continuous change in phase as it slowly shows more of its lit side after its new phase until we see the full Moon. It then shows less and less of its lit side until it gets back to new phase after 29.5 days. The selection of only six photos for this experience rather than eight is to reinforce the idea that there are not eight specific lunar phases. The photos are also oriented in a random pattern to encourage students to look at the features on the Moon to determine "which way is up" when viewing each image. This provides a reason for the students to learn about craters, maria, and rays. For your reference only at this time, the correct order of the images is in Figure 4.15 (p. 299).

MATERIALS

One per group:

- "Six Lunar Photographs, Set 1" (p. 291)

- Blank sheet of paper

- Scissors

- Tape or glue

One per student:

- Astronomy lab notebook

4.1

ENGAGE

Advance Preparation

Make a copy of the "Six Lunar Photographs, Set 1" handout for each group.

Procedure

1. Tell the students that a teacher colleague of yours (identify by name if desired) had a set of lunar photographs sent to him or her by an amateur astronomer friend. Unfortunately, your colleague dropped the photos on the floor and no longer knows the order or orientation of the photographs. The friend asked for help from you and your class to get them in the correct order. You have made copies of the photos so the students can help with this challenge.

2. Distribute copies of the "Six Lunar Photographs, Set 1" handout, scissors, tape or glue, and a blank sheet of paper to each work group. It is important to use a photocopy machine that will preserve the detail in the photos. The students' goal is to place the photographs on the sheet of paper in the order in which they think they would see the shape of the Moon if they made observations for several weeks. Allow 10 to 15 minutes for discussion and decision making in each group.

3. Once each group is satisfied with the order, students should tape or glue the photos to the blank sheet of paper. Have them number the pictures from 1 to 6 in the order each would be seen. Be sure they indicate which way is up. They should also put their names on the paper to show who made the prediction.

4. When all of the groups have completed their photo sequences, have them move around the room to see the predictions of other groups. Ask the work groups, one by one, to explain their reasoning for choosing the sequence they came up with. These reasons should not yet be judged for appropriateness since the students are only presenting their best guess.

5. Use this discussion as a transition to Experience 4.3, "Observing the Moon," by pointing out that the best way to know the correct order is to observe the Moon over a number of days. The students' predictions should be posted on a wall of the room for ongoing reference during Experience 4.3. Alternatively, one

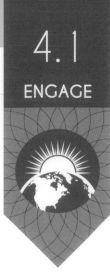

member of each work group can keep the team's photo sheet in his or her astronomy lab notebook for later reference.

Teacher note: Students will want to be immediately be told the "right" answer for the order of the Moon photographs. It is important not to share the right answer at this point but to use Experience 4.3 as a way for students to discover the correct order for themselves.

6. The discussion in step 4 is also an effective transition to the next *engage* experience, whichlets students think about and express what they already know about lunar phases and eclipses.

Six Lunar Photographs, Set 1

Source:
Fred Espenak,
www.astropixels.com

EXPERIENCE 4.2

What Do We Think We Know?

Overall Concept

Just as in prior chapters, this "What Do We Think We Know?" experience allows students to reflect on what they already know about a topic covered in this chapter. This gives you an understanding of what knowledge they already have and if they have any preconceptions that need to be dealt with.

Objectives

Students will

1. reflect on and write in their astronomy lab notebooks what they think they know about lunar phases and solar and lunar eclipses; and

2. share their ideas with other students.

Advance Preparation

Identify space where the list of student preconceptions can be located for the length of time the class is researching lunar phases and eclipses.

Procedure

1. Tell students that now that they have explored the lunar images in Experience 4.1, "Predicting How the Moon Will Look," and have a prediction for what they will see when they observe the Moon, you would like to understand what they already know about the Moon's phases and related phenomena.

2. Have them label the top of a page in their astronomy lab notebook with "What I Think I Know." Right below that, have them write the

MATERIALS

For the class:

- Space on a whiteboard, blackboard, or piece of poster paper where student preconceptions can be recorded and kept visible for the duration of the time researching lunar phases and eclipses

One per student:

- Astronomy lab notebook

first question you want them to consider and discuss: "What are the different phases of the Moon, and how do the phases change throughout the month and year?" Explain, if necessary, that "phases" are the shape that the Moon appears to have when we look at it in the sky. At the top of the following page, have them write the next question: "What causes the different phases?" On the third page, ask them to write the final prompt: "What are solar and lunar eclipses, and what causes them?"

3. Use the think-pair-share process described in the book's Introduction (p. xix) to have students (1) individually write answers to the questions in their notebooks; (2) discuss their answers with other students in small groups and add more detail to their notebooks as desired; and finally, (3) report out in their groups and write a list of what they, as a whole class, think they know in a location in the classroom where the information can be kept for future reference.

4. Finish up the experience by thanking the students for sharing what they think they know and emphasizing that keeping the answers up will help the class see what new things they learn during the study of the Moon's phases and its relationship to lunar and solar eclipses.

5. As you do the experiences in this chapter, you should periodically refer back to what students said in this experience to see how their thinking has changed.

EXPERIENCE 4.3

Observing the Moon

Overall Concept

A major outcome of Experience 4.1, "Predicting How the Moon Will Look," is that students want to know who has the right sequence for the phases of the Moon, so they are motivated to go outside to observe the Moon, which is the focus of this experience. To completely identify the appropriate sequence— and orientation—of the photos, the students need to be able to identify a number of features on the lunar surface, so this activity also allows for a study of lunar craters and maria. (Just a reminder that the word maria means "seas" in Latin and is plural; the singular is mare.)

Objectives

Students will

1. make a daily record of their Moon observations;

2. identify features (e.g., craters, maria) that they see on the lunar surface; and

3. use their observations to develop an understanding of the sequence of lunar phases and the location of a select number of lunar features.

Advance Preparation

Experience 4.1, "Predicting How the Moon Will Look," provides an excellent introduction to this experience, so we suggest you

MATERIALS

For the class:

- A large poster or projection of the "Lunar Map" (p. 300)

- A large poster of the "Lunar Observing Record Chart" (p. 301)

One per group:

- "Six Lunar Photographs, Set 1" (from Experience 4.1, p. 291)

- Predictions of the order of the lunar phases from Experience 4.1, if you did that experience

One per student:

- "Lunar Map" (p. 300)

begin this experience at its conclusion. This experience is ideally started a few days before the Moon is at first quarter. The Moon will be in the western sky in the afternoon and evening, which will allow you to take students outside near the end of the school day to make the first observation together. Some students may not realize that the Moon is often visible in the daytime as well as at night, so you may want to have the students think about the different times of the day they have seen the Moon. This first daytime observation allows you to review with the students what each lunar observation consists of and gets them into the routine of making daily observations either when the entire class can observe together or on their own. With your assistance, students will then be able to use their skills to make nighttime observations in the coming weeks when the Moon is not visible in the daytime sky.

You may also find it useful to provide the students with a chart (see Table 4.5) that tells them when the Moon will be above the horizon for some of the key phases.

If you have multiple classes or cannot take students out near the end of the day to make the observations, you can ask them to do their first Moon observation as homework on the way home or shortly after they get home.

You can easily find when the Moon is near first quarter by looking on a calendar or searching the internet for "phases of the Moon." You can start the experience at any time, but certain phases work better for

MATERIALS (*continued*)

- "Lunar Observing Record Chart" (p. 301); you may provide two charts per student if you want them to make observations over a longer period of time

- Pencil

- Clipboard or other firm writing surface

- Astronomy lab notebook

- (*Optional*) Binoculars—some schools may have extra binoculars to lend to students overnight on a rotating basis

TABLE 4.5

Rising and setting times of the Moon during key phases

Moon phase		Approximate rise time	Approximate set time
New Moon	●	Sunrise	Sunset
First quarter	◐	Noon	Midnight
Full Moon	○	Sunset	Sunrise
Third quarter	◑	Midnight	Noon

making observations during times when students are outside or awake. You should also check the weather reports to help identify a day around first quarter when it is likely to be clear in the afternoon.

We highly recommend that you do this experience yourself a month or two in advance of doing it with the class so that you will be prepared for some of the challenges that students will encounter—primarily bad weather, trying to observe from a location where buildings or trees block the view of the Moon, or looking at the wrong time of day.

Procedure

1. Distribute the "Lunar Map" handouts and have students use the maps to identify some key features they should look for when out observing the Moon. These features should be large or noticeable for another reason (e.g., different in coloring), so students can see them with no equipment except their eyes. Ask the students to find the features on the photographs they sequenced in Experience 4.1. This is also a good time to provide more information about craters, maria, rays, and what caused them. See the Content Background section on the surface features of the Moon (pp. 277–279) for more information about these features.

2. Distribute copies of the "Lunar Observing Record Chart." Tell the students they will now have an opportunity to observe the Moon themselves to determine the correct order for the lunar photographs. They will also be able to explore some of the ideas they raised in Experience 4.2, "What Do We Think We Know?" (if you did that experience). This is also a good time to introduce the astronomical vocabulary regarding what we call each phase of the Moon (Figure 4.14).

    Teacher note: Ideally, the students will make observations over an entire month. If time constraints or the weather do not allow the students to observe the Moon for that long, they should be able to begin determining the pattern of the phases after about 10 observations. Observations are not required every day, so some days without observations should be fine. When the weather becomes a problem (clouds for more than two or three days in a row), students can use internet resources to complete their observation charts.

FIGURE 4.14

Names of the phases of the Moon

| New Moon | Full Moon | Waxing crescent | Waning crescent |

| Third quarter | First quarter | Waning gibbous | Waxing gibbous |

Websites for finding the current lunar phase include the following:

- Calculator Cat: *www.calculatorcat.com/moon_phases/phasenow.php*

- Calendars Through the Ages. *www.webexhibits.org/calendars/moon.html*

- Moonpage.com: *www.moonpage.com*

3. Explain how the "Lunar Observing Record Chart" is used. This is best done by going outside with the class to locate and make the first observation of the Moon together. Bring a pair of binoculars with you if you have one so that students can get a better look at some of the surface features, which are often difficult to see in the daytime. Record the date, the time of the observation, the Moon's location in the sky, and its shape. Add small drawings that show the shape and location of lunar features that can be identified from the lunar map.

Teacher note: When the Moon is in a crescent or quarter phase, it can often be difficult to be sure which feature is being seen. Tell the students to make their best guess and then follow the feature during the coming days as they make observations.

The key features will become obvious as the Moon approaches full Moon.

4. Have students go out every clear day or night to repeat their observations. Encourage them to use binoculars if they have access to a pair to help identify various lunar surface features. After the first observation, make a class activity of predicting what phase the Moon will be in before the next observation.

    Teacher note: In some urban areas, parents may not be comfortable with students going outside at night to find the Moon. You may want to send a sheet about this experience home with students and get a sense of how parents feel and whether they would be comfortable going out with their students to help. If not, then students could use internet resources to fill in the phases for days when the Moon is only visible at night. Also, binoculars may not be available to students at home. If the school has binoculars in quantity, you might want to arrange for a loan program, or you may want to make the identification of features at home a less important part of this experience.

    It is helpful to summarize daily observations on a classroom copy of the "Lunar Observing Record Chart" that can be posted on a wall. An alternative to this is to have one student draw a picture each day on construction paper of the class's observations for the previous day. These could be posted daily in consecutive order to allow the students to see the pattern of the phases emerge throughout the activity. After a number of days, students should be encouraged to compare their daily observations to the predictions they made in Experience 4.1. Some students will want to start making changes to their predictions. Tell them that they will have plenty of opportunity to compare their observations to their predictions, but for now they need to leave their predictions unaltered.

5. After observing the Moon for 10 to 30 days (which are required to see the pattern for the lunar phase cycle), it is time to summarize what the students learned from their observations. Use the think-pair-share process to have students write and discuss what their observations revealed about the phases of the Moon. The following are key ideas that should emerge, assuming you are observing from typical Northern latitudes in the continental United States:

4.3

EXPLORE

a.   The phases started with a crescent Moon that had sunlight on the right side (assuming you started a few days before first quarter).

b.   More and more of the Moon facing the Earth became lit by the Sun over the next week or two until it was all in sunlight (full Moon).

c.   After the full Moon, less and less of the Moon facing the Earth was lit by the Sun, and the lit part was on the left side.

d.   If students observed for a full month, then they should be able to conclude that the time to go from a given phase back to that phase is approximately a month (actually, it's 29.5 days).

e.   Although the amount of light on the Moon's surface facing toward the Earth changes throughout the month, the features on the Moon appear to stay in the same location on the side of the Moon we can see.

This is a good time to talk about the relationship between the lunar phase cycle and why we divide the year into months. You might also discuss that many cultures had calendars based on the lunar cycle rather than the cycle of the Earth's orbit around the Sun.

6.   Conclude the experience by giving each group another copy of the "Six Lunar Photographs, Set 1" handout and asking them to redo the ordering process. This new order should be attached to the original paper just below their initial predictions. Once they have done this, you can confirm the appropriate order for the photographs (see Figure 4.15) and also review the different phases the Moon goes through. Finally, you can reinforce that the Moon takes 29.5 days to go through a full set of phases.

7.   If you need to assess individual student understanding of lunar phases, this is an appropriate time do the first *evaluate* experience (Experience 4.10, "Lunar Phases Revisited").

FIGURE 4.15

Correct order for the lunar photographs in set 1

1. Waxing crescent

2. First quarter

3. Waxing gibbous

4. Full Moon

5. Waning gibbous

6. Waning crescent

Source:
Fred Espenak,
www.astropixels.com

Lunar Map

Source:
Fred Espenak,
www.astropixels.com

Lunar Observing Record Chart

Sunday	Monday	Tuesday	Wednesday	Thursday	Friday	Saturday
◯ Date___ Time___ Location:	◯ Date___ Time___ Location:	◯ Date___ Time___ Location:	◯ Date___ Time___ Location:	◯ Date___ Time___ Location:	◯ Date___ Time___ Location:	◯ Date___ Time___ Location:
◯ Date___ Time___ Location:	◯ Date___ Time___ Location:	◯ Date___ Time___ Location:	◯ Date___ Time___ Location:	◯ Date___ Time___ Location:	◯ Date___ Time___ Location:	◯ Date___ Time___ Location:

Three-Dimensional Learning Exposed

Three-dimensional learning (the integration of the disciplinary core ideas, science practices, and crosscutting concepts from the *Next Generation Science Standards*) is best demonstrated in this chapter when combining Experience 4.1, "Predicting How the Moon Will Look," Experience 4.3, "Observing the Moon," and Experience 4.4 "Modeling the Moon."

Not only do students develop an understanding of the "cyclic patterns of lunar phases [and] eclipses," as written in the middle school performance expectation MS-ESS1-1, but they engage in the following key science practices:

- Analyze and interpret data during their efforts to predict the order of the lunar phases and then when they make the regular observations of the Moon in the sky.

- Use a model of the Earth–Moon–Sun system (lightbulb, Styrofoam balls, and their heads) to describe the relationship between the Earth, Moon, and Sun—thus developing an understanding of what causes the Moon's phases and both lunar and solar eclipses.

- Engage in argument from evidence as they compare their predictions for the order of the lunar photographs and during their daily observations of the Moon.

These experiences also allow the teacher to identify crosscutting concepts embedded in the learning. The key crosscutting concepts include the following:

- Patterns observed in the experiences can identify cause-and-effect relationships, as seen in how the relative position of the Earth, Moon, and Sun are the cause of the Moon's phases.

3D
LE

- Science assumes that objects and events in natural systems occur in consistent patterns that are understandable through measurement and observation, as demonstrated by observations of the Moon and Sun leading to an understanding of when solar and lunar eclipses occur, allowing astronomers to predict when future eclipses will be visible.

- System models provide an opportunity for understanding and testing ideas, as provided in the model of the Earth-Moon-Sun system created using the student's head, Styrofoam ball, and lightbulb.

As mentioned in Chapter 1, we suggest you take time during the experiences to emphasize the value of the science practices and crosscutting concepts in all areas of science.

EXPERIENCE 4.4

Modeling the Moon

Overall Concept

Now that students understand the order of the lunar phases and the length of the cycle, the typical question they bring up is, "What causes the phases?" This experience allows students to understand the cause by building on the modeling activity from Chapter 1 in which the students' heads were the Earth and a lightbulb at the front of the room was the Sun. The Moon is now added to the model—in the form of a small Styrofoam ball attached to a pencil. This allows the students to explore the relationships among the Earth, Moon, and Sun to understand what causes lunar phases.

Objectives

Students will

1. be able to identify the order of the Moon's phases from one full Moon to the next; and

2. demonstrate how the Moon's position around the Earth (relative to the Sun) creates the phases.

Advance Preparation

Be sure there is plenty of space for students to stand with a hand stretched out and to spin around as they work through this experience. Check that the lamp or lightbulb for the model Sun works properly and can be placed high in the front of the room for everyone to see it. The room you use

MATERIALS

For the class:

- Lightbulb on a stand or clamp (or a lamp with its shade removed); a 60 W bulb is best for this experience

- Extension cord

- A room that can be made completely dark

One per student:

- Smoothfoam or Styrofoam ball or light-colored sphere for each student (as a model Moon)

  Teacher note: Smoothfoam balls have a denser, smoother surface that works better for this activity, but they are often harder to find than Styrofoam balls. Either will work. Places on the web that sell Smoothfoam or Styrofoam balls include Michaels (*www.michaels.com*) or Smoothfoam.com (*www.smoothfoam. com/category/balls.html*). Staples and other similar companies also carry them online.

- Pencil

- "Moon Phase Activity Sheet" (p. 310)

- Astronomy lab notebook

The Sun, the Moon, and the Earth Together: Phases, Eclipses, and More

for this experience needs to be completely dark, which often means you have to switch rooms or spend time putting up black plastic sheets, dark tablecloths, or poster boards to cover light leaks in your classroom.

Procedure

1. Review the results of Experience 4.3, "Observing the Moon," which showed that the Moon goes through a sequence of phases. Work with the students to review the order of the phases from one full Moon to the next. Discuss some of the students' predictions about what causes the lunar phases, if this was explored in earlier discussions.

2. Tell the students that since we cannot go to outer space to observe the Moon orbiting Earth and the change in phase, we will use a model to learn what causes the Moon phases. Make the room completely dark and place the lamp at the front. Remind students of safety near the hot lightbulb and electrical cord. Have students stand facing the lamp. Make sure they are spread out enough that light from the lamp reaches each student. If you did Experience 1.5, you can remind students that this activity will be an extension of their model Sun–Earth system. The lamp still represents the Sun and their heads still represent Earth, with their noses being the students' hometown.

3. Review what they learned from the model of the Earth and Sun developed in Chapter 1. Ask students to stand so it is noon in their hometown (noses-at-noon). If disagreement occurs as to what position this would be, have students discuss until it is agreed that noon is when their nose is pointed toward the model Sun. Next, ask them to stand so it is midnight at their noses. They should turn so that they face away from the Sun.

4. Students should recall which way Earth rotates on its axis from the experiences in Chapter 1. If students did not do those experiences or do not remember them, you will need to review a few things. Determine which way north, south, east, and west are for their model Earths (their heads). If their hometown (nose) is in the Northern Hemisphere, north is the top of their heads, south is their chins, east is to their left, and west is to their right. From prior knowledge and their Moon observations, they should know

4.4

EXPLAIN

that the Sun rises in the east. Have the students place their open hands on the sides of their heads, acting as horizon blinders. Have them determine which way Earth rotates so that the Sun rises over their left (eastern) hand. After some trial and error, they will be able to determine that the Earth rotates from right to left in their model, with their right shoulder moving forward (Figure 4.16).

FIGURE 4.16

Diagram showing how the students should stand in the model

5. Ask students to stand so it is sunrise and then so it is sunset. Practice the ideas of sunrise, noon, midnight, and sunset until you feel that the students have a good understanding of these relative positions. This is a good review of what they learned in Chapter 1, and it gives them some practice with the model before introducing the Moon.

6. Distribute one "Moon ball" to each student. Have them stick a pencil into the ball to make it easier to hold and observe the phases of the Moon in the model. If there is already a hole in the ball from previous use, tell them to use that one and not make a new one. Have students hold the model Moon at arm's length. Allow time for students to explore how the Sun's light reflects off the model as they place their Moons in different positions around the Earth. This is a good time to tell students that the Moon orbits the Earth in a counterclockwise direction when

looking down on the Earth and Moon from above the North Pole. As they explore the different lunar phases, remind them to always have the Moon move in the correct direction.

One question that usually comes up and must be addressed is how high the model Moon should be held. If it is held at head height, there will be an eclipse (instead of a full Moon) during each orbit of the Moon around the student's head. Help the students develop the idea that they did not observe a lunar eclipse during Experience 4.3, and generally people make a big deal about eclipses. Therefore, they probably do not occur every month. Students should then conclude that they have to hold the Moon balls up high so the balls are exposed to the Sun's light throughout their orbit around Earth. The topic of eclipses is covered in Experience 4.5, "Modeling Eclipses," and in Experience 4.6, "How Often Do Eclipses Occur?" In Experience 4.6, they will learn that the Moon's orbit is not aligned with the Earth's orbit around the Sun (or relative to the circle that the Sun appears to make among the constellations in the course of a year, which is how we on Earth perceive our motion around the Sun). As a result, the Moon is usually either above or below the Sun in the sky. But if you plan to do Experience 4.6, you may not want to give away the answer while you help them with the current activity.

7. Help students find a few of the phases of the Moon with which they are already familiar, such as a full Moon, a new Moon, and the first and third quarters. A new Moon occurs when the Earth, Moon, and Sun are aligned, and the Moon is between Earth and the Sun. A full Moon occurs when the three bodies are aligned, and the Earth is between the Moon and the Sun (Figure 4.17, p. 308).

Teacher note: Students will have many questions as they explore. Try not to answer directly. Encourage them to explore their questions using the model before providing an answer.

There is a common misconception that Earth's shadow causes the phases, and some of your students may try to involve the shadow of their heads in the modeling. If students are trying to produce the different phases by hiding parts of the Moon with shadows of their heads, you will need to address this. Students should also come to recognize, possibly with some assistance, that they cannot generate the shape of the different phases by using the shadow of Earth.

8. After students explore finding the phases, choose one lunar phase and ask the students to determine what position in the Moon's orbit they must place their Moon to achieve that particular phase. Full Moon is a good phase to start with. Encourage students to compare their positions and discuss differences. Ask a student who has the correct position to explain why it is correct. By walking around the classroom, you can check for understanding by seeing if all the students are holding their Moons in the same position.

9. Have students model other phases, for example, first quarter, third quarter, and new Moon. Use the terminology introduced in Experience 4.3 when requesting a particular phase, such as waning gibbous and third quarter.

FIGURE 4.17

A diagram of the Moon phases in relation to the Sun and Earth

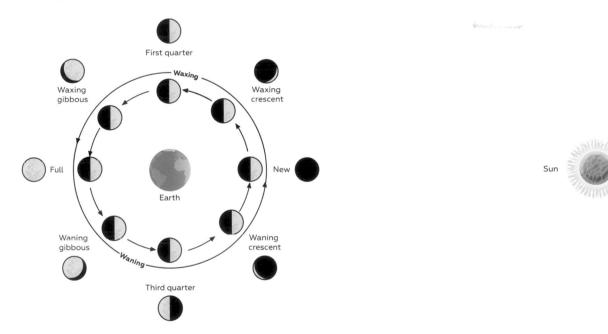

The inner sequence shows the Moon's relative position to the Earth and the Sun as viewed from outer space, above the solar system. Students are asked to produce a portion of this diagram on the "Moon Phase Activity Sheet." The outer sequence shows the Moon as seen from Earth. For example, you would see the waning crescent (lower right) as a small slice of the Moon illuminated on the left side. A waxing crescent, upper right, would have the right side of the Moon illuminated.

10. Allow time for students to experiment with the movement of the Moon—always moving it in a counterclockwise direction around the Earth. They can observe their own model as well as other students' models. This activity is very powerful and can answer many questions that the students generate about the motion of the Moon and its appearance in the sky.

    Teacher note: Students may find it helpful to change the model slightly to answer certain questions. If one student holds the Moon ball and another student is Earth, they can more easily see Earth spinning on its axis while the Moon is barely moving in its orbit. How much of a circle does the Moon travel each 24 hours? (*About 1/29th or 1/30th of a circle.*) So everyone on Earth basically sees the same phase on the same night.

11. Now have students work together in small groups as they each complete the "Moon Phases Activity Sheet." The goal is for them to produce a diagram similar to the one in Figure 4.17. These drawings should be kept in their astronomy lab notebooks.

12. After completing the diagrams, ask students to write down in their astronomy lab notebook the causes of the changing Moon phases. (*The movement of the Moon around Earth and the relative positions of the Sun, Earth, and Moon cause the phases. The spinning Earth—the student's head—makes the Moon rise and set each day, but this does not affect the phase of the Moon.*) Encourage them to use diagrams in their explanations.

13. Check student diagrams and explanations for the causes of phases. Ask students if they are sure that their observations and the model support their diagrams and statements. If discrepancies arise, have students go back to the model to further clarify the concepts.

    Teacher note: If you have not already used the first *evaluate* experience (Experience 4.10, "Lunar Phases Revisited"), now is a good time to do Experience 4.10 to assess student understanding of lunar phases.

Moon Phase Activity Sheet

This diagram represents a view you would see looking down from above at your head when you are modeling the Moon orbiting Earth. Darken the areas on each Moon that are not illuminated by the Sun. Then label each Moon phase as you would see it when your nose (on Earth) is pointed directly at it.

Be sure to use the Moon phase terms: new Moon, full Moon, first quarter, third quarter, waxing crescent, waning crescent, waxing gibbous, and waning gibbous.

Phase _____

Phase _____

Phase _____

Phase _____

Phase _____

Phase _____

Phase _____

Phase _____

Earth

Sun

NATIONAL SCIENCE TEACHERS ASSOCIATION

EXPERIENCE 4.5

Modeling Eclipses

Overall Concept

Now that students understand where the Moon has to be in its orbit to see each phase, modeling continues by exploring where the Moon, Earth, and Sun have to be to produce solar and lunar eclipses.

Objectives

Students will

1. distinguish between lunar and solar eclipses and

2. understand that solar eclipses only happen when the Moon is in its new phase and lunar eclipses only occur when it's in its full phase.

Procedure

1. If you did Experience 4.2, "What Do We Think We Know?" ask students to review their response to prompt 3, "What are solar and lunar eclipses, and what causes them?" If you did not do Experience 4.2, ask students if they know what an eclipse is. What is the difference between a solar eclipse and a lunar eclipse? Ask them if they have ever seen an eclipse, and if so, whether it was solar or lunar. Have them write responses to the questions in their astronomy lab notebooks. Then explain that this activity will help them understand the difference between these two types of eclipses as well as why they occur.

2. Set up the equipment in the same way as it was used in Experience 4.4, "Modeling the Moon." Have the students explore moving their model Moon in its orbit (always revolving counterclockwise) to determine when the Moon can block the Sun's light from reaching the Earth (a

MATERIALS

For the class:

- Lightbulb on a stand or clamp (or lamp with the shade removed)

- Extension cord

One per student:

- Smoothfoam or Styrofoam ball or light-colored sphere (as a model Moon)

- Pencil

- Blank sheet of paper

- Astronomy lab notebook

solar eclipse) and when the Earth can block the Sun's light from hitting the Moon (a lunar eclipse). If you recently did Experience 4.4, in which you told the students that the shadow of the Earth was not a factor, you may want to tell them that shadows are a factor for eclipses, and they can now feel free to play with shadows. As much as possible, have them use the model to come up with the correct answer. (*Solar eclipses only occur when the Moon is in its new phase. Lunar eclipses only occur when the Moon is in its full phase.*)

3. After the students have explored how to produce the two types of eclipses, reinforce what they observed by having them move the Moon ball in orbit until it completely blocks their view of the lamp. Explain that it is when the Moon is positioned exactly between the Earth and the Sun that it blocks the Sun and produces a solar eclipse. Ask them what phase the Moon must be in to line up and make a solar eclipse. (*New Moon*) This is a good time to show several photographs of different solar eclipses. These are easily found by searching the internet for solar eclipse images.

4. Now ask them to position their Moon balls so that the Sun's light falling on the Moon is blocked by the Earth. Ask them what phase the Moon must be in to produce this lunar eclipse. (*Full Moon*) This is a good time to show several photographs of different lunar eclipses.

5. Have the students take notes in their astronomy lab notebooks and briefly write the reason that we have eclipses. They should explain what is necessary for us to experience a solar eclipse and a lunar eclipse. Ask them to explain what the two types of eclipses have in common. (*The Sun, Earth, and Moon are all lined up.*)

    Teacher note: A question that typically arises at this point is why we don't have solar and lunar eclipses every month since we have a new Moon and full Moon every month. The Moon's orbit around the Earth is out of alignment by 5° from the plane of Earth's orbit around the Sun (the ecliptic). This means the Moon is not lined up with the Earth and the Sun most months. This is the subject of *elaborate* Experience 4.6, "How Often Do Eclipses Occur?" which makes a great extension for this experience.

EXPERIENCE 4.6

How Often Do Eclipses Occur?

Overall Concept

Students continue to use the model Sun, Earth, and Moon to explore the answer to this question. At the conclusion of this experience, Hula-Hoops are used to model the Moon's orbit around the Earth and the Earth's orbit around the Sun to help students understand why there are typically two solar eclipses and two lunar eclipses each year.

Objectives

Students will understand that

1. the orbital planes of the Moon going around the Earth and of the Earth going around the Sun are slightly inclined relative to each other; and

2. this inclination results in there only being eclipses when the orbits intersect and in there being in general pairs of solar and lunar eclipses occurring twice during the year, about six months apart.

Advance Preparation

Before doing this experience, students should have done Experience 4.5, "Modeling Eclipses," so they start the experience knowing that solar eclipses occur when the Moon is in its new phase and passes in front of the Sun from the viewpoint of a person on the Earth. Similarly, they should know that a lunar eclipse occurs when the Moon is in its full phase and passes into the shadow of the Earth.

MATERIALS

For the class:

*   Lightbulb and Styrofoam ball setup from prior experiences in this chapter

*   Dark room

*   2 Hula-Hoops (it helps to wrap one in yellow tape and the other in gray tape)

One per student:

*   Astronomy lab notebook

4.6
ELABORATE

Procedure

1. Remind the students that they discovered in Experience 4.5 that the Moon has to be full for us to see a lunar eclipse and has to be in its new phase for us to see a solar eclipse. But that experience also raised the question regarding why these eclipses do not occur every month. With the room dark and the Sun lightbulb on, ask them to experiment moving their model Moons in their orbit as they generate ideas regarding why eclipses don't occur each month and how often they think eclipses will occur. Encourage them to write these possibilities in their astronomy lab notebooks. After a period of exploration, have students explain their ideas using arguments based on the model. This will typically lead to someone suggesting that sometimes the Moon passes slightly above or below the Sun or the Earth's shadow. It will typically be difficult for them to predict how often eclipses should occur.

2. Once you've had enough discussion, bring out the Hula-Hoops and tell the students that these can help them determine the answer to the question. Hold two Hula-Hoops over your head, as shown in Figure 4.18, to demonstrate the relationship between the paths of the Sun and Moon as seen from Earth (your head).

3. The inside hoop is the orbit of the Moon, with a full cycle of phases happening every 29.5 days. Have one student use their Moon ball to follow the path of the Moon around the Earth. Discuss the route the Moon takes as it orbits the Earth.

4. The outer hoop represents the Sun's path as seen from Earth, with the Sun appearing to go around the Earth once a year. Have one student trace the Sun's path around the hoop. Ask

FIGURE 4.18

Two Hula-Hoops can be used to show the relationship between the path of the Sun and the path of the Moon around Earth.

how it differs from the Moon's path. (*The two paths are slightly inclined relative to each other.*)

5. Ask students where the Moon and Sun need to be so that they are perfectly in line. (*At the crossing points of the orbits.*) Since the Sun is only at these points twice a year, this determines how often we have eclipses. When the Sun is at one of the crossing points, a solar and lunar eclipse will generally occur within two weeks of each other—the time it takes for the Moon to get from new Moon orientation to full Moon orientation, or vice versa. Roughly six months later, there will usually be another solar and lunar eclipse separated by two weeks.

6. Have students do a post-experience writing assignment in their astronomy lab notebooks. They should now be able to accurately define the two types of eclipses as well as explain how often eclipses occur and why. They should us diagrams to help in their explanations.

EXPERIENCE 4.7

Why Do People Spend $10,000 to See a Total Solar Eclipse?

Overall Concept

Another question that often comes up from students is, "Why do people travel long distances and spend lots of money to see a solar eclipse, but no one takes a trip to see a lunar eclipse?" Students use the model Earth–Moon–Sun to discover that the shadow of the Moon only covers a small portion of the Earth (the student's head) during a solar eclipse, leading to the realization that people who want to see a solar eclipse often need to travel to a distant location on Earth. Students then use the model to discover that this is not true for lunar eclipses, which all the people on the half of the Earth facing the Moon get to see.

Objective

Students will understand why more people on the Earth can see a total lunar eclipse than a total solar eclipse.

Procedure

1.   A good way to start this experience is to find an ad on the internet for a tour to go see a total solar eclipse in a distant location. Tell the students that people who read astronomy magazines regularly see ads like this to go see solar eclipses but never to see a lunar eclipse. Say, "Let's see if we can figure out why." With the room dark and the Sun lightbulb on, ask them to experiment by moving their model Moon in their orbit as they generate ideas regarding why people will "shell out" large sums of money to see a solar eclipse. Encourage them to work in their small groups to come up with possible reasons to write in their astronomy lab notebooks. After a period of exploration, have students explain their ideas using arguments based on the model.

MATERIALS

For the class:

*   Lightbulb and Styrofoam ball setup from prior experiences in this chapter

*   Dark room

One per student:

*   Astronomy lab notebook

2.   After completing a discussion of the possibilities, have the students work in pairs with their model Sun, Moon, and Earth. One partner holds the Moon ball an arm's length away from the other student, so it produces a solar eclipse on the other person's head (the Earth). The person holding the Moon ball should look at the shadow of the Moon falling on the face of his or her partner. If the student's head were the Earth, from what part of the Earth could people see the solar eclipse? (*Just the small area where the Moon ball's shadow falls on the partner's face.*) Have the partner then hold the ball so there is a lunar eclipse (i.e., the ball is now in the shadow of your partner's head). Ask if more people will see a lunar eclipse or a solar eclipse. (*Everyone on the side of the Earth facing toward the Moon will see the lunar eclipse.*) Have the partners change places and repeat the observations.

3.   Students should now be able to conclude—and you can summarize—that many more people on the Earth can see a lunar eclipse than can see a solar eclipse. Only those people in the narrow path where the Styrofoam ball (the Moon) made a dark shadow on the student's head (the Earth) will see the total solar eclipse, whereas people on the entire half of the Earth facing the Moon will see the lunar eclipse.

4.   This is a good time to also add that at any point on the Earth, the total eclipse of the Sun typically lasts only a few minutes before the shadow of the Moon moves on. An eclipse of the Moon, on the other hand, typically takes several hours. The Earth's shadow slowly darkens the Moon. The Moon then stays darkened for an extended period of time and finally slowly moves out of the Earth's shadow, which allows the full light of the Sun to again shine on the face of the Moon. Another reason that total solar eclipses are so spectacular to see is that the sky goes dark during the day as the Sun is covered (and the stars actually come out). During the total lunar eclipse, it is already dark, and darkening the Moon only makes it a bit darker. People are much more surprised and awed by the change in the light during a total solar eclipse.

5.   Have students do a post-experience writing assignment in their astronomy lab notebooks. They should now be able to accurately explain why fewer people will see a total solar eclipse than a total lunar eclipse. They should include diagrams to help in their explanations.

EXPERIENCE 4.8

Does the Moon Rotate?

Overall Concept

During Experience 4.3, "Observing the Moon," students discover that we always see the same features on the side of the Moon facing toward Earth and realize that it must keep the same face toward us. This reinforces a major preconception that many people have—that the Moon must not rotate. This experience uses a model in which students' heads are the Moon and the Earth to demonstrate that the Moon does rotate. What is special about the Moon's rotation period is that it is the same as the period of its revolution around the Earth. This makes the same side of the Moon always face toward the Earth.

Objectives

Students will

1. model the two motions of the Moon as it orbits Earth and spins on its axis; and

2. realize that the Moon rotates at the same rate that it orbits the Earth, which is why we always see the same lunar features facing toward Earth.

Procedure

1. Tell students that you have had a number of people over the years ask you whether the Moon rotates or not, and you want to see what they say about this question. Have students use the think-pair-share process to first write their own ideas in their astronomy lab notebooks, next discuss their ideas with a small group of classmates, and then have a general discussion with the entire class. This is not the time to come to a conclusion but simply to generate ideas.

MATERIALS

For the class:

*   A large, clear area for students to move around

One per pair of students:

*   Styrofoam Moon ball

*   Colored pushpin or tack

One per student:

*   Astronomy lab notebook

2. Have students work in pairs, with one person being the Earth and the other person holding the Moon ball as they orbit the Earth and rotate the Moon ball at the same time. Have them use the pushpin or tack to mark a spot on the Moon to represent a large crater. Students should work together, trying different ways the Moon can orbit and spin, to see if they can build evidence for whether the Moon rotates or not. When they think they know the answer, have them switch places and repeat their observations. They should record their observations and reasoning in their astronomy lab notebooks.

3. After 10 to 15 minutes, ask them to share their conclusion with a couple of other pairs. Then have a general discussion with the entire class. Be sure that students always back up their arguments with observations they made with their Moon balls.

4. Typically, there may still be disagreement on whether or not the Moon rotates, so ask the students to participate in one more modeling activity. Start this model by asking the students to stand up and slowly turn around once, observing the parts of the room they see as they spin. They will see all four corners of the room and all the other students in the room. Conclude that they see all points in the room as they spin.

5. Have the students get back into pairs. Remind students that in Experience 4.3 they saw the same features on the Moon throughout the month. One student will be the Earth (his or her head) and the other Moon (his or her head). Ask the Moon to orbit the Earth, always keeping his or her face toward Earth. Thus, the person representing the Earth always sees the Moon's face. Once students have practiced this a few times, have the Moon-student watch the view of the room as they orbit Earth, while always facing toward the Earth. They will see all four corners in the room, just as they did when they spun around earlier. This means that as the Moon orbits the Earth, it must slowly spin on its axis so that the same face is always toward the Earth. Ask Moon-students how many times they see each corner of the room during one circle around the Earth-student. (They should conclude the answer is once.) Put another way, the Moon-student will be turning in a circle at the same rate that he or she

moves around the Earth-student. This shows that the Moon's rotation rate is the same as its revolution around the Earth.

6. Now have the Moon-students not spin as they orbit. The best way to do this is for each student to notice what wall in the room he or she is facing. They should continue to face that direction as they orbit the Earth-student. The Earth-student should comment on what part of the Moon-student's head they see, especially as the Moon-student moves each quarter of the way around the Earth-student. The Earth-student will report that he or she sees all parts of the other student's head, not just his or her face.

7. Have students do a post-experience writing assignment in their astronomy lab notebooks. They should now be able to accurately explain how we know the Moon rotates. They should include diagrams to help in their explanations.

What Do Eclipses Look Like From a Space Colony on the Moon?

4.9

ELABORATE

Overall Concept

Students use their Earth–Moon–Sun model to predict what they would see during a solar eclipse if they were astronauts at a Moon base. This provides students with the opportunity to use the Earth–Moon–Sun model under new conditions, thus deepening their skill in using and understanding models.

Objectives

Students will

1. deepen their experience with and understanding of the Earth–Moon–Sun model used in this chapter and

2. understand what a person located at a lunar base would see during a solar and a lunar eclipse.

Procedure

1. A good way to start this experience is to ask how many students would like to live in or visit a Moon colony when they are built in the future. Use the ensuing discussion to bring up the following question: Suppose there is an eclipse visible on Earth while you are visiting the Moon colony. What would the colonists see when people on Earth see a lunar or a solar eclipse?

2. Use the think-pair-share process to have groups of students consider what they would see, asking them to think about the activities they did using the Earth–Moon–Sun model. If they want, they can make sketches in their astronomy lab notebooks, but don't give them Moon balls to use for modeling yet. This gets them to think more abstractly about the question before you introduce the physical model in step 4.

MATERIALS

For the class:

- Lightbulb and Styrofoam ball setup from prior experiences in this chapter

- Dark room

One per student:

- Astronomy lab notebook

3. Have a whole-class discussion of the various student predictions after the students do some individual writing in their astronomy lab notebooks and have small group discussions. Save the results of the whole-class discussion for use with the follow-up discussion at the end of the experience.

4. Now have the students pair up and explore answering this same question using the model Earth, Moon, and Sun. One student should be the Earth while the other pretends to be a colonist on the Moon, keeping the Styrofoam ball near his or her face as the Moon orbits the other student's head (Earth). This is a good time to reinforce the concept that the Moon must be in new phase (between the Earth and Sun) for the Earth-based viewer during a solar eclipse and must be in full phase (the Moon on the other side of the Earth from the Sun) for the Earth-based viewer during a lunar eclipse.

5. Have the student pairs work as independently as possible, and give your attention to those students needing the most help. Be sure the students record their observations and conclusion in their astronomy lab notebooks. After 5 to 10 minutes, have them switch roles and repeat the observations.

6. After another 5 to 10 minutes, have them share their observations and conclusion with one or two other pairs of students. Have them continue to add notes to their astronomy lab notebooks.

7. Have a whole-class discussion to reach a general consensus—or as much of a consensus as you can get—on what a lunar colonist would see. If necessary, you can ask two students to "act out" the relationship of the Earth and Moon while you point out what a colonist would see.

   Teacher note: Typically, students assume that the colony is on the side of the Moon that always faces toward the Earth. In this case, a colonist would see the following:

   a. Total solar eclipse: As the Moon passes in front of the Sun, the shadow of the Moon that falls on the Earth causing the total eclipse will be a spot of dark traveling from west to east across the Earth. Lunar colonists would see an Earth with a shadow spot on it. This is a good time to reinforce that this path of totality is generally less than 160 mi. wide and a person on the Earth is generally in the Moon's shadow for less than five minutes.

b.   Total lunar eclipse: As the Moon passes into the shadow of the Earth, the portion of the Sun covered by the Earth as seen from the Moon increases until the Earth totally covers the Sun. Lunar colonists would see an eclipse of the Sun by the Earth. The Sun eventually reemerges from the other side of the Earth as the Moon continues in its orbit. The Sun typically stays hidden behind the Earth for an hour or less, the same amount of time a person on the Earth would see the total lunar eclipse. A reddish halo would be seen around the edge of the Earth because of the sunlight refracted by the Earth's atmosphere. This reddish light is what travels out to the Moon and is then reflected back to Earth, giving the Moon a reddish glow during a total lunar eclipse.

If students have not brought up the alternative of what would be seen if the colony were on the far side of the Moon, you might want to raise this question and leave it for the students to consider on their own time. If the colony is on the far side, then the colonists would see the following:

a.   Solar eclipse: The colonists are on the side facing the Sun and facing away from the Earth, so no Earth is visible in the sky. It would be a very sunny day at the colony, and the colonists would not see any trace of the eclipse occurring on the Earth.

b.   Lunar eclipse: The colonists on the far side of the Moon are experiencing nighttime since the Sun is shining on the other side (the one that faces the Earth, which is what produces a full Moon on Earth). Neither the Earth nor Sun would be in the lunar sky on the far side, so these colonists would not see an eclipse.

8.   Conclude the experience by asking interested students to search the internet for images of eclipses as seen from the International Space Station or simulations of eclipses as seen from the Moon and then share them with the rest of the class.

   Teacher note: Such images can be found at *http://apod.nasa. gov/apod/ap990830.html, http://earthobservatory.nasa.gov/ IOTD/view.php?id=6419,* and *http://phys.org/news/ 2012-05-eclipse.html.*

EXPERIENCE 4.10

Lunar Phases Revisited

Overall Concept

Students receive a different set of six photographs of the Moon than were used in Experience 4.1, "Predicting How the Moon Will Look," and are asked to repeat the process placing the images in the order in which they think they would see the shape of the Moon if they made observations over a month's time.

Objectives

Student will demonstrate how well they

1. know the order of the lunar phases and

2. can orient the photographs based on lunar surface features.

Advance Preparation

Make a copy of the "Six Lunar Photographs, Set 2" handout for each student.

Procedure

> *Teacher note:* The students already ordered the first set of lunar photographs twice, so this experience is to see how well they can apply what they know to a different set of lunar images. This experience can also be used to assess the understanding of individual students, if that is needed.

1. Distribute copies of the "Six Lunar Photographs, Set 2" handout, scissors, tape or glue, and a blank sheet of paper (if they are not doing this in their notebooks) to each student. Ask them to place the photographs on the sheet of paper in the order in which they would see the shape of the Moon if they made observations for several weeks.

MATERIALS

One per student:

*   "Six Lunar Photographs, Set 2" (p. 326)

*   Scissors

*   Tape or glue

*   Blank piece of paper or blank page in each student's astronomy lab notebook

2. Once the student is satisfied with the order of the photos, each student should tape or glue the photos to a page in his or her notebook or on the blank sheet of paper. They should number the pictures from 1 to 6 in the order each would be seen and indicate which way is up. Encourage them to look back at their previous work during Experiences 4.1 and 4.3 to help them determine the appropriate order and orientation for the photographs. Figure 4.19 shows the correct order for the photographs. The order starts with a waxing crescent, but the sequence could start with any phase.

3. At the bottom of the page, ask students to write a justification for the order they chose. Ask them to include details regarding the following:

   a. Specific lunar features (e.g., craters, maria) that helped them orient the photos

   b. Different lunar phases and the location of the dark versus lit portions of the Moon that helped them order the photos

FIGURE 4.19

Correct order for the six lunar photographs from set 2

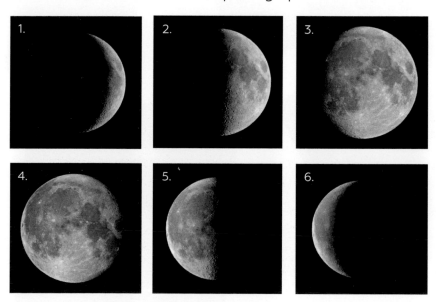

Six Lunar Photographs, Set 2

Source:
Fred Espenak,
www.astropixels.com

EXPERIENCE 4.11

What Causes Lunar Phases and Eclipses?

Overall Concept

Students label a diagram that shows the Earth, Sun, and Moon to indicate where the Moon is in its orbit to produce each phase and to show where the Moon must be in its orbit to produce solar and lunar eclipses.

Objective

Students will demonstrate how well they understand that the relative position of the Moon, Earth, and Sun produces lunar phases and solar and lunar eclipses.

Procedure

Teacher note: These same instructions are on the worksheet for students to read as they complete the experience. Have students do the following:

1. Cut off the top half of the "Causes of Lunar Phases and Eclipses Diagram" along the dotted line and attach it to a blank page in their astronomy lab notebook.

2. Cut out the eight images of the Moon on the bottom half of the worksheet and attach each one at an appropriate place on the top half of the worksheet.

3. Label the appropriate phases for the Moon when people on the Earth can see a lunar or a solar eclipse.

4. At the bottom of the page, explain what causes the lunar phases and eclipses. Ask them to include explanations regarding these specific items:

MATERIALS

One per student:

*   "Causes of Lunar Phases and Eclipses Diagram" (p. 330)

*   Astronomy lab notebook

a. Explain the way the light from the Sun falls on the
    Moon and how this produces the phases we see. Include
    detailed explanations for at least two different phases
    on your diagram.

b. Why can't solar and lunar eclipses occur at other
    phases than the ones you indicate?

Causes of Lunar Phases and Eclipses Diagram

TEACHER VERSION*

The full Moon phase should have an arrow (or other identification method) pointing to it with a label that indicates this is the phase when lunar eclipses can occur. The new Moon phase should have an arrow (or other identification method) pointing to it with a label that indicates that this is the phase when solar eclipses can occur.

The explanation at the bottom of the page needs to include these key components to be fully correct:

1. The side of the Moon toward the Sun is always lit while the other side is dark.

2. Depending on where the Moon is in its orbit around us, we on Earth only see a portion of the lit side, which produces the different phases (e.g., at new Moon, the back side of the Moon is lit, so we see a dark Moon; at first quarter, the right side of the Moon is lit and the left side is in the dark, so we only see the right half of the Moon facing the Earth in sunlight).

3. Eclipses cannot occur at any other phase than new and full because the Earth, Moon, and Sun must be lined up to create an eclipse. This only occurs at new and full phases.

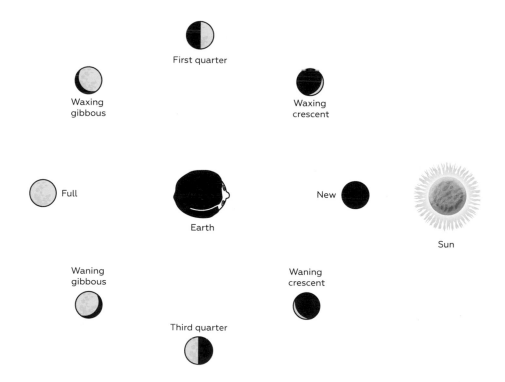

*A copy of the student handout is available at www.nsta.org/solarscience.

Causes of Lunar Phases
and Eclipses Diagram

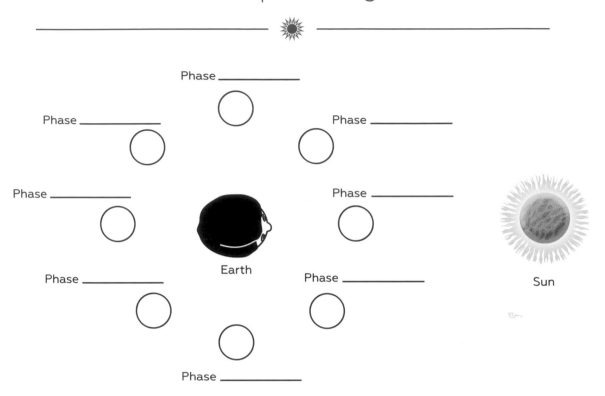

Phase _____

Phase _____

Phase _____

Phase _____

Phase _____

Earth

Phase _____

Phase _____

Phase _____

Phase _____

Sun

1. Cut off the top half of the "Causes of Lunar Phases and Eclipses Diagram" along the dotted line and attach it to a blank page in your astronomy lab notebook.

2. Cut out each of the eight Moon phase images on the bottom half of the worksheet and attach each one at an appropriate place on the diagram, showing where the Moon must be located in its orbit to produce that phase.

3. Label the appropriate phase for the Moon when people on the Earth can see a lunar or a solar eclipse.

4. At the bottom of the page, explain what causes the lunar phases and eclipses. Be sure to include explanations regarding these specific items:

   a. Explain the way the light from the Sun falls on the Moon and how this produces the phases we see. Include detailed explanations for at least two different phases on your diagram.

   b. Answer the question, "Why can't solar and lunar eclipses occur at other phases than the ones you indicate?"

Video Connections

- *Mr. Lee's Phases of the Moon Rap*. A rap song to help students remember the phases and their cause (4 min.): *www.youtube. com/watch?v=79M2lSVZiY4*

- *Crash Course Astronomy*. If you would like to see the directly opposite technique from ours for teaching about the Moon, check out the Moon phases episode of *Crash Course Astronomy* with Phil Plait, in which the phases are explained as an illustrated lecture at breakneck speed (10 min.): *www.youtube.com/ watch?v=AQ5vty8f9Xc*

- *Understanding Lunar Eclipses*. This video by NASA Goddard explains the reason why there isn't an eclipse every month with good animation (2 min.): *www.youtube. com/watch?v=lNi5UFpales*

- *Shadow of the Moon*. This NASA Goddard video explains eclipses of the Sun, with discussion and animation, focusing on the 2015 eclipse and showing what an eclipse looks like from space (2 min.): *www. youtube.com/watch?v=XNcfKUJwnjM*

- *Solar Eclipse: "Breathtaking" Views Witnessed by Millions*. A BBC News report on the 2015 total and partial eclipse of the Sun that gives a nice sense of people enjoying the event and an animation to show what's happening (3 min.): *www. youtube.com/watch?v=-PhjA6nstvU*

- *Solar Eclipse Maths and the Cosmic Coincidence of the Saros Cycle*. A comedy-style lecture by Matt Parker, stand-up mathematician, explaining the alignments that make eclipses possible and the 18 year Saros cycle of repeating eclipse geometries (14 min.): *www.youtube.com/ watch?v=ieUvzy6rnnw*

- *Physical Science 9.2a: The Earth Moon Sun System*. A short video demonstrating scale models of the distances between the Earth and the Moon and the Earth and the Sun (7 min.): *www.youtube.com/ watch?v=FjCKwkJfg6Y*

Math Connections

- NASA has a free book on math about the Moon, called *Lunar Math*, that includes a few problems about phases and lunar cycles of time (see sections 2 and 3). You can find the whole book free at *www.nasa. gov/pdf/377727main_Lunar_Math.pdf*.

- Modern astronomers use the term "blue Moon" to mean the second full Moon in the same month. Ask students to think about which months can have two full Moons and which can't. If a month has a blue Moon in it, can the next month have a blue Moon too? Explain.

- If you look at the table of future eclipses in this chapter, it seems that the timing and location of eclipses around the world doesn't have a pattern to it (other than the pattern of "eclipse seasons" roughly six months apart). However, astronomers have discovered that there are long-scale patterns in the occurrence of eclipses. If you have students who like playing with numbers, basic statistics, and patterns, they may want to read more about eclipse cycles. See, for example, the NASA Eclipse

Web Site, which is maintained by Fred Espenak at *http://eclipse.gsfc.nasa.gov/ SEsaros/SEperiodicity.html*.

- NASA Connect has an activity booklet called *Path of Totality: Measuring Angular Size and Distance* available at *www. knowitall.org/nasa/pdf/connect/Path_ of_Totality.pdf*. In this activity, students come to understand how the Sun, which is bigger but farther away, can look to be the same size in our sky as the Moon, which is smaller but closer. Students are then asked to see if the coincidence of the Sun and Moon being the same angular size would be true for other planet-moon combinations in our solar system.

Literacy Connections

- Two websites by Mr. Eclipse (NASA astronomer Fred Espenak):

    - Espenak, F. 2009. Lunar eclipses for beginners. Mr. Eclipse.com. *www. mreclipse.com/Special/LEprimer. html*.

    - Espenak, F. 2009. Solar eclipses for beginners. Mr. Eclipse.com. *www. mreclipse.com/Special/SEprimer. html*.

- Soderman, T., and NASA Lunar Science Institute. How are craters formed? NASA Solar System Exploration Institute. *http://sservi.nasa.gov/articles/ how-are-craters-formed*.

- Phillips, T. 2001. The great Moon hoax: Moon rocks and common sense prove

Apollo astronauts really did visit the Moon. Science@NASA. *http://science.nasa.gov/ science-news/science-at-nasa/2001/ ast23feb_2*.

Cross-Curricular Connections

1. Writing prompt about Moon phases: Give students the assignment to write a short story in which the plot involves the phases of the Moon. For example, a rescue or theft can only be carried out when it is completely dark at night, so the people involved must wait for the next new Moon. Or a special party or event requires lots of light at night, so everyone agrees that it must be held on the full Moon. Alternatively, the story might require light before midnight but not after—ask the students what phase the Moon must be in to provide that timing.

2. Eclipses that affected history: Both solar and lunar eclipses have affected historical events over the years. You might assign students (individually or in groups) to research some of these historical stories and write them up or present them to the entire class.

    Here are some resources for learning more:

Book and Articles

- Olson, D. W. 2014. *Celestial sleuth: Using astronomy to solve mysteries in art, history and literature*. New York: Springer-Praxis.

- Olson, D. W. 1992. Columbus and an eclipse of the Moon. *Sky and Telescope* (October): 437.

- Schaefer, B. 1992. Lunar eclipses that changed the world. *Sky and Telescope* (December): 639.

- Schaefer, B. 1994. Solar eclipses that changed the world. *Sky and Telescope* (May): 36.

Websites

- Espenak, F. Solar eclipses of historical interest. NASA. *http://eclipse.gsfc.nasa. gov/SEhistory/SEhistory.html.*

- Espenak, F. Lunar eclipses of historical interest. NASA. *http://eclipse.gsfc.nasa. gov/LEhistory/LEhistory.html.*

3. Eclipses on stamps: Eclipses make deep impressions on people in all walks of life. When a country is in the path of a total eclipse, there is often a push to commemorate the occasion with stamps. Any students who are stamp collectors, or who have a collector in the family, may want to find some eclipse stamps from different countries and research the eclipse that each stamp commemorates. (See *www. mreclipse.com/SEstamps/SEstamps1. html* and *mseclipse.free.fr/timbres/ timbres.htm* for examples.)

4. Eclipses in legends and mythology: Many older cultures tried to celebrate and explain eclipses through stories they told. Students can research the eclipse mythology of one or more cultures and discuss how the stories cover some of the properties of eclipses that they learned about in class. See, for example, the following articles:

- Wanner, N. n.d. The Sun-eating dragon: Eclipse stories, myths, and legends.

Exploratorium. *www.exploratorium.edu/ eclipse/dragon.html.*

- An interview with astronomer and mythology expert E.C. Krupp:
  Lee, J. J. 2013. Solar eclipse myths from around the world. National Geographic. *http://news.nationalgeographic.com/ news/2013/11/131101-solar-eclipse-myth- legend-space-science.*

- A similar article about lunar eclipse myths:
  Lee, J. J. 2014. Lunar eclipse myths from around the world. National Geographic. *http://news.nationalgeographic.com/ news/2014/04/140413-total-lunar- eclipse-myths-space-culture-science.*

5. Fiction that involves eclipses: Eclipses play a role in many stories, novels, TV shows, and films. You could ask your students if they ever read a story or saw a show or film with an eclipse. If not, they might ask adults in their lives for suggestions. Alternatively, they can do some research on the internet. See, for example, the Wikipedia article "Solar Eclipses in Fiction" at *http://en.wikipedia.org/wiki/ Solar_eclipses_in_fiction.*

   One classic novel in which an eclipse plays a role is Mark Twain's *A Connecticut Yankee in King Arthur's Court* (1889). The crucial scene can be found at the NSTA web site for this book at *www.nsta.org/solarscience.*

6. Music about eclipses: Students who have an interest in music might want to look for songs or classical music that involve or were inspired by eclipses of the Sun and the Moon. Among those we have found are the following:

- "Total Eclipse" by George F. Handel is an aria from the oratorio *Samson* (many recordings). The poignant song compares Samson going blind with an eclipse of the Sun.

- *Sonata on the Long Eclipse of the Moon July 6, 1982*, by Alan Hovhaness (Nicola Giosmin on piano, produced by TauKay), portrays some of the physical and personal aspects of watching the full Moon turn dark and then come back into the light. Hovhaness is a prolific 20th century Armenian-American composer who plays with astronomy in a number of his pieces.

- *Notes on Light* by Kaija Saariaho (performed by Orchestre De Paris, produced by Ondine) is a modern piece for cello and orchestra in which the composer tries to represent properties of light and phenomena with light through musical textures. The fourth movement is "Eclipse."

- "Eclipse" by GeorgeTsontakis (piece for clarinet, violin, cello, and piano; produced by Broyhill Chamber Ensemble on Koch) was written after the composer observed a lunar eclipse in 1995. The music reminds him of "the eclipsing shadow… as it softly invaded the hazy luminescent circle, and later, the shadow leaving the sphere just as quietly as it had first entered."

- "Total Eclipse of the Heart" by Bonnie Tyler on *Faster Than the Speed of Night* (produced by Sony) is a 1983 song by a Welsh singer using eclipse images— shadows, being in the dark, "no one in the universe as magical as you"—to describe a love affair going wrong.

Resources for Teachers

ABOUT THE PHASES OF THE MOON

- Great images and technical information about the phases of the Moon can be found on former NASA scientist Fred Espenak's photography page:
  Espenak, F. Astropixels. *http://astropixels.com/moon/phases/phasesgallery.html*.

- NASA's Space Place also has a good sequence of Moon phase images:
  NASA. The Moon's phases in Oreos. *http://spaceplace.nasa.gov/oreo-moon/en/Moon_phases_all_L.en.jpg*.

- Current Moon phase calculators can be found at the following sites:

  - Star Date. Moon phases. *http://stardate.org/nightsky/Moon*.

  - Calculator Cat. Current Moon phase. *www.calculatorcat.com/Moon_phases/phasenow.php*.

  - Moonpage. *www.moonpage.com/index.html*.

  - Walker, J. Earth and Moon viewer. Fourmilab. *www.fourmilab.ch/earthview/vplanet.html*. (See the second paragraph for a variety of current views of the Moon.)

ABOUT ECLIPSES

BOOKS AND ARTICLES

- Primer on the astronomy and the observation of eclipses, by three experienced astronomy authors:
  Littmann, M., F. Espenak, and K. Willcox. 2009. *Totality: Eclipses of the Sun.* 3rd ed. New York: Oxford University Press.

- A guide from a veteran amateur astronomer and author:
  Harrington, P. 1999. *Eclipse! The what, where, when, why and how guide to watching solar and lunar eclipses.* New York: John Wiley and Sons.

- A historical and observing guide, translated from French:
  Brunier, S., and J.-P. Luminet. 2000. *Glorious eclipses: Their present, past, and future,* trans. S. Dunlop. Cambridge, U.K.: Cambridge University Press.

- An introductory book with quite a bit of history by an astronomer author:
  Steele, D. 2001. *Eclipse: The celestial phenomenon that changed the course of history.* Washington, DC: Joseph Henry Press.

- This article describes the circumstances of upcoming total eclipses of the Sun:
  Bakich, M. 2008. Your twenty-year solar eclipse planner. *Astronomy* (October): 74.

- A look at what we have learned and are still learning from eclipses:
  Pasachoff, J. 2001. Solar eclipse science: Still going strong. *Sky and Telescope* (February): 40.

- Information on eclipse chasing as a hobby:
  Regas, D. 2012. The quest for totality. *Sky and Telescope* (July): 36.

WEBSITES

- NASA astronomer Fred Espenak earned the nickname Mr. Eclipse for his tireless work calculating and explaining eclipses. This site includes introductory and explanatory material on eclipses in general and upcoming eclipses:

  - Espenak, F. Total eclipse of the Moon. Mr. Eclipse. *www.mreclipse.com/MrEclipse.html.*

  - Espenak, F. Solar eclipse preview, 2011–2030. Mr. Eclipse. *www.mreclipse.com/Special/SEnext.html.*

  - Espenak, F. Lunar eclipse preview, 2011–2030. Mr. Eclipse. *www.mreclipse.com/Special/LEnext.html*

- *Sky and Telescope* magazine hosts a page about eclipses that contains links to articles on upcoming eclipses and eclipse observing and photography tips:
  Sky and Telescope. Eclipses. *www.skyandtelescope.com/observing/objects/eclipses.*

- NASA's eclipse website is *the* site for reliable information on eclipse paths and circumstances in the past and the future; the site is a bit technical, so beginners should start with the Mr. Eclipse site above:
  NASA eclipse web site. NASA. *http://eclipse.gsfc.nasa.gov/eclipse.html.*

- Jay Anderson has created a site devoted to weather and maps for upcoming eclipses:

Anderson, J. Eclipser: Climatology and maps for the eclipse chaser. *http://home.cc.umanitoba.ca/~jander.*

ABOUT OBSERVING THE MOON FOR BEGINNERS

- An introduction to observing the Moon from *Sky and Telescope* magazine:
  MacRobert, A. 2014. Take a Moon walk tonight. Sky and Telescope. *www.skyandtelescope.com/observing/take-a-Moon-walk-tonight.*

- An introductory article and video from *Astronomy Magazine*:
  Bakich, M. Observing the Moon through binoculars. Astronomy Magazine. *www.astronomy.com/observing/observe-the-solar-system/2014/06/observe-the-Moon-through-binoculars.*

- Advice for getting the most out of Moon obvservations:
  Bakich, M. E. 2010. Ten tips for Moon watchers. *Astronomy Magazine* (February): 52. *www.astronomy.com/~/media/Files/PDF/web%20extras/2011/10/10%20tips%20for%20Moon%20watchers.pdf.*

- Help with using modest telescopes for observing and photography:
  Burnham, R. and G. Therin. 1991. The joys of Moongazing. *Astronomy Magazine* (March): 84.

- Hints for finding the Moon as soon after its "new Moon" phase as possible:
  MacRobert, A. and R. Sinnott. 2005. Young Moon hunting. *Sky and Telescope* (February): 75.

OTHER TOPICS IN THE CHAPTER

- A NASA lesson on synchronous rotation:
  NASA. What is synchronous rotation? *Educational Brief. http://saturn.jpl.nasa.gov/files/Synchronous_Rotation.pdf.*

For more resources on the lunar calendar, see Chapter 2.

Image Credits

p. vi: ThinkStock

p. x: Association of Science-Technology Centers

p. xi: Tucker Hiatt

Introduction

Figure I.1: Photo by Senior Airman Joshua Strang, U.S. Air Force, Wikimedia Commons, Public domain. *https://commons.wikimedia.org/wiki/File:Polarlicht_2.jpg*

Figure I.2: ThinkStock

Figure I.3: NASA/Goddard Space Flight Center/Conceptual Image Lab

Figure I.4: : Astronomical Society of the Pacific

Figure I.5: Galileo Galilei, public domain. *https://scienceofthestars.files.wordpress.com/2013/04/gallileo_moon2.jpg*

p xxi: ThinkStock

Chapter 1

Figure 1.1: Joe Butera

Figure 1.2: Photograph copyright Danilo Pivato

Figure 1.3: Joe Butera

Figure 1.4: (a) Dennis Schatz; (b) ThinkStock; (c) ThinkStock

Figure 1.5: User:Tentotwo, Wikimedia Commons, CC BY-SA 3.0. *https://commons.wikimedia.org/wiki/File:Magnetic_North_Pole_Positions.svg*

p. 12: ThinkStock

Figure 1.6: Schatz, D., and P. Allen. 2003. A*stro adventures II: An activity-based astronomy curriculum.* Seattle, WA: Pacific Science Center, p. 30. Reprinted with permission.

Figure 1.7: Schatz, D., and P. Allen. 2003. *Astro adventures II: An activity-based astronomy curriculum.* Seattle, WA: Pacific Science Center, p. 33. Reprinted with permission.

Figure 1.8: Schatz, D., and P. Allen. 2003. A*stro adventures II: An activity-based astronomy curriculum.* Seattle, WA: Pacific Science Center, p. 19. Reprinted with permission.

p. 29: Schatz, D., and P. Allen. 2003. A*stro adventures II: An activity-based astronomy curriculum.*Seattle, WA: Pacific Science Center, p. 21. Reprinted with permission.

Figure 1.9: Schatz, D., and P. Allen. 2003. A*stro adventures II: An activity-based astronomy curriculum.* Seattle, WA: Pacific Science Center, p. 27. Reprinted with permission.

Figure 1.10: Joe Butera

Figure 1.11: Joe Butera

Figure 1.12: Dennis Schatz

Figure 1.13: Dennis Schatz

Figure 1.14: Schatz, D., and P. Allen. 2003. A*stro adventures II: An activity-based astronomy curriculum.* Seattle, WA: Pacific Science Center, p. 48. Reprinted with permission.

p. 46: ThinkStock / Joe Butera

p. 65: ThinkStock

Chapter 2

Figure 2.1: NASA

Figure 2.2: Joe Butera

Figure 2.3: Joe Butera

p. 78: ThinkStock

Figure 2.4: User:Creysmon07, Wikimedia Commons, CC BY-SA 2.0. *https://commons.wikimedia.org/wiki/File:Prospect_Heights_Blizzard_NYC_2-12-06.jpg*

Figure 2.5: Dennis Schatz

Figure 2.6: Dennis Schatz

Figure 2.7: Schatz, D., and P. Allen. 2003. A*stro adventures II: An activity-based astronomy curriculum.* Seattle, WA: Pacific Science Center, p. 94. Reprinted with permission.

p. 103: Schatz, D., and P. Allen. 2003. A*stro adventures II: An activity-based astronomy curriculum.* Seattle, WA: Pacific Science Center, p. 102. Reprinted with permission. Modifed by Joe Butera.

p. 106: Schatz, D., and P. Allen. 2003. A*stro adventures II: An activity-based astronomy curriculum.* Seattle, WA: Pacific Science Center, p. 105. Reprinted with permission.

p. 108: Schatz, D., and P. Allen. 2003. A*stro adventures II: An activity-based astronomy curriculum.* Seattle, WA: Pacific Science Center, p. 107. Reprinted with permission.

Figure 2.8: Schatz, D., and P. Allen. 2003. A*stro adventures II: An activity-based astronomy curriculum.* Seattle, WA: Pacific Science Center, p. 94. Reprinted with permission.

p. 111: Schatz, D. and P. Allen. 2003. *Astro adventures II: An activity-based astronomy curriculum.* Seattle, WA: Pacific Science Center, p. 109. Reprinted with permission.

p. 112: Schatz, D. and P. Allen. 2003. *Astro adventures II: An activity-based astronomy curriculum.* Seattle, WA: Pacific Science Center, p. 94.

Figure 2.9: Joe Butera

Figure 2.10: Dennis Schatz

Figure 2.11: Dennis Schatz

Figure 2.12: Joe Butera

Figure 2.13: Joe Butera

pp. 144, 145: Joe Butera

p. 153: NASA Earth Observatory image by Robert Simmon

Chapter 3

Figure 3.1: NASA

Figure 3.2: National Oceanic and Atmospheric Administration

Figure 3.3: European Space Agency

Figure 3.4: NASA, SOHO mission image

Figure 3.5: NASA

Figure 3.6: Joe Butera

Figure 3.7: Simo Räsänen, Wikimedia Commons, CC BY-SA 3.0. *https://commons.wikimedia.org/wiki/File:Aurora_borealis_above_Lyngenfjorden,_2012_March.jpg*

Figure 3.8: Photo sequence by Peter van de Haar; used with permission. *http://www.footootjes.nl/Astrophotography_Planets_Stars/20111003_Marching_sunspots_processed_880x1675.jpg*

Figure 3.9: Lemuel Francis Abbott, National Portrait Gallery; Wikimedia Commons, Public domain. *https://commons.wikimedia.org/wiki/File:William_Herschel01.jpg.*

p. 172: NASA

Figure 3.10: NASA

p. 177: NASA

Figure 3.11: Schatz, D., and P. Allen. 2003. *Astro adventures II: An activity based astronomy curriculum*. p. 52. Seattle, WA: Pacific Science Center.

Figure 3.12: NASA

Figure 3.13: David Hathaway, NASA Marshall Space Flight Center, Wikimedia Commons, Public domain. *https://commons.wikimedia.org/wiki/File:Solar_Cycle_Prediction.gif*

p. 189: NASA

p. 194: Illustration by Galileo Galilei from the Galileo Project. *http://es.rice.edu/newgalileo/images/things/sunspot_drawings/ss623-l.gif*;

p. 195: National Solar Observatory

Figure 3.14: NASA

Figure 3.15: NASA

Figure 3.16: Joe Butera

pp. 204, 205: NASA

Figure 3.17: NASA (a and b)

p. 211: NASA

p. 213: Joe Butera

pp. 214, 215: NASA

Figure 3.18: Lars Tiede, Wikimedia Commons, CC BY-SA 2.0. *https://commons.wikimedia.org/wiki/File:Northern_lights_on_Kval%C3%B8ya_2012-01-23a.jpg*

Figure 3.19: Ben Hider / The Central Intelligence Agency, Wikimedia Commons, Public domain. *https://commons.wikimedia.org/wiki/*

File:Director_Petraeus_rings_opening_bell_at_NY_Stock_Exchange_-_Flickr_-_The_Central_Intelligence_Agency.jpg

Figure 3.20: Jet Propulsion Laboratory

Figure 3.21: Albert Londe, Wikimedia Commons, Public domain. *https://commons.wikimedia.org/wiki/File:X-ray_1896_nouvelle_iconographie_de_salpetriere.jpg*

Figure 3.22: NASA

p. 242: NASA

pp. 253–255: Joe Butera

p. 256: ThinkStock

p. 262: David Hathaway, NASA

Chapter 4

Figure 4.1: NASA/JPL

Figure 4.2: Schatz, D., and P. Allen. 2003. *Astro adventures II: An activity-based astronomy curriculum*. Seattle, WA: Pacific Science Center, p. 138. Reprinted with permission.

Figure 4.3: Fred Espenak, *www.astropixels.com*

Figure 4.4: Fred Espenak, *www.astropixels.com* / Joe Butera

Figure 4.5: NASA/JPL/USGS

Figure 4.6: Joe Butera

Figure 4.7: Joe Butera

Figure 4.8: (a) Kevin Baird, Wikimedia Commons, CC BY-SA 3.0. *https://commons.wikimedia.org/wiki/File:Annular_eclipse_%22ring_of_fire%22.jpg*. (b) Luc Viatour, Wikimedia Commons, CC BY-SA

3.0. *https://en.wikipedia.org/wiki/File:Solar_ eclipse_1999_4_NR.jpg.* (c) NASA.

Figure 4.9: Xavier Jubier. Used with permission.

Figure 4.10 Joe Butera

Figure 4.11: Javier Sánchez, Wikimedia Commons, CC BY-SA 3.0. *https://commons.wikimedia.org/wiki/ File:Eclipse_lunar_total.jpg*

Figure 4.12: Joe Butera

Figure 4.13: Joe Butera

p. 291: Fred Espenak, *www.astropixels.com*

Figure 4.14: Schatz, D., and P. Allen. 2003. A*stro adventures II: An activity-based astronomy curriculum.* Seattle, WA: Pacific Science Center, p. 128. Reprinted with permission.

Figure 4.15: Fred Espenak, *www.astropixels.com*

p. 300: Fred Espenak, *www.astropixels.com*

p. 301: Schatz, D., and P. Allen. 2003. A*stro adventures II: An activity-based astronomy curriculum.* Seattle, WA: Pacific Science Center, p. 129. Reprinted with permission.

Figure 4.16: Schatz, D., and P. Allen. 2003. A*stro adventures II: An activity-based astronomy curriculum.* Seattle, WA: Pacific Science Center, p. 27. Reprinted with permission.

Figure 4.17: Schatz, D., and P. Allen. 2003. A*stro adventures II: An activity-based astronomy curriculum.* Seattle, WA: Pacific Science Center, p. 138. Reprinted with permission.

p. 310: Schatz, D., and P. Allen. 2003. A*stro adventures II: An activity-based astronomy curriculum.* Seattle, WA: Pacific Science Center, p. 141. Reprinted with permission. Modified by Joe Butera.

Figure 4.18: Schatz, D., and P. Allen. 2003. A*stro adventures II: An activity-based astronomy curriculum.* Seattle, WA: Pacific Science Center, p. 165. Reprinted with permission.

Figure 4.19: Fred Espenak, *www.astropixels.com*

p. 326: Fred Espenak, *www.astropixels.com*

p. 329: Schatz, D., and P. Allen. 2003. A*stro adventures II: An activity-based astronomy curriculum.* Seattle, WA: Pacific Science Center, p. 165. Reprinted with permission. Modified by Joe Butera.

p. 330: Schatz, D., and P. Allen. 2003. A*stro adventures II: An activity-based astronomy curriculum.* Seattle, WA: Pacific Science Center, pp. 128 and 141. Reprinted with permission.

Index

Page numbers printed in **boldface** type refer to tables or figures.

NATIONAL SCIENCE TEACHERS ASSOCIATION